An Introduction to Interdisciplinary Research

An introduction to Interdisciplinary Research

2nd revised edition

Machiel Keestra

Anne Uilhoorn

Jelle Zandveld

Amsterdam University Press

Volume 2 of the Series Perspectives on Interdisciplinarity

Cover design and lay-out: Matterhorn Amsterdam

ISBN   978 94 6372 469 2
e-ISBN   978 90 4855 759 2
DOI   10.5117/9789463724692
NUR   143

© M. Keestra, A.G. Uilhoorn, J. Zandveld / Amsterdam University Press B.V., Amsterdam 2022

All rights reserved. Without limiting the rights under copyright reserved above, no part of this book may be reproduced, stored in or introduced into a retrieval system, or transmitted, in any form or by any means (electronic, mechanical, photocopying, recording or otherwise) without the written permission of both the copyright owner and the author of the book.

Every effort has been made to obtain permission to use all copyrighted illustrations reproduced in this book. Nonetheless, whosoever believes to have rights to this material is advised to contact the publisher.

Contents

	Acknowledgments		**8**
	Preface		**11**
	Chapter guide		12
Part 1	**The Handbook – 'The What'**		**14**
1	**Introduction**		**15**
2	**What is science? A brief philosophy of science**		**21**
	2.1	What is science?	21
	2.2	Moving through the science cycle	24
3	**From disciplines to interdisciplinarity**		**42**
	3.1	Knowledge formation and the emergence of disciplines	42
	3.2	Pre-disciplinary approaches to the organization of knowledge	43
	3.3	From knowledge formation to disciplines	45
	3.4	Discipline: a plural concept	46
	3.5	Increasing specialization and isolation of disciplines	48
	3.6	Unification of science or collaborations between disciplines	50
4	**Interdisciplinarity and its ongoing developments**		**52**
	4.1	Weaving disciplines: multidisciplinarity, interdisciplinarity, and transdisciplinarity	52
	4.2	Interdisciplinarity: a definition and an explanation	54
	4.3	Variations in interdisciplinary research	55
	4.4	Drivers of interdisciplinarity	59
	4.5	Complexity and complex adaptive systems	62
	4.6	Wicked problems	65
	4.7	Transdisciplinary research	66
	4.8	Action research	69
5	**Interdisciplinary integration**		**71**
	5.1	Main categories of the interdisciplinary integration toolbox	75
	5.2	Final remarks on integration	89

Part 2		The Manual – 'The How'	**90**
	6	**The interdisciplinary research process**	**91**
	6.1	The interdisciplinary research model	91
	6.2	Interdisciplinary collaboration in research	95
		Exercises	98
	7	**The problem**	**99**
		Step 1 Identify problem or topic	99
		Step 2 Formulate preliminary research question	101
		Exercises	103
	8	**Theoretical framework and research question**	**105**
		Step 3 Identifying relevant theories for each discipline	107
		Step 4 Identifying overlap and distinctions between different perspectives	112
		Step 5 Finalize interdisciplinary research question and sub-questions	119
		Exercises	120
	9	**Data collection and analysis**	**121**
		Step 6 Determine research methods and design	122
		Step 7 Collect and analyze data	126
		Exercises	128
	10	**Completion of your research**	**129**
		Step 8 Integrate results related to the sub-questions	130
		Step 9 Interpret results, discuss research, draw conclusions	130
		Exercises	133

Part 3	**Interdisciplinary research in practice**		**134**
	11	**Interdisciplinary research example**	**135**
		Step 1 Determine the problem or topic	135
		Step 2 Formulate preliminary interdisciplinary research question	136
		Step 3 Identify relevant theories for each discipline	136
		Step 4 Identify overlap and distinctions between different perspectives	137
		Step 5 Finalize interdisciplinary research question and sub-questions	138
		Step 6 Determine research methods and design	139
		Step 7 Collect and analyze data per sub-question	140
		Step 8 Integrate results related to the sub-questions	140
		Step 9 Interpret results, discuss research, and draw conclusion(s)	141
	12	**Interdisciplinary careers**	**142**

Further reading — 150

Index — 156

References — 158
    References Research projects undertaken by IIS students — 166

Colophon — 167

Acknowledgments

This handbook is based on almost 20 years of experience with interdisciplinary studies accumulated by the authors, editors, and others involved in writing this book. The approach we have adopted relies heavily on the materials and practices used by the Institute for Interdisciplinary Studies (IIS) at the University of Amsterdam (UvA). A first edition of this handbook, edited by Menken & Keestra and written by Rutting, Post, Keestra, de Roo, Blad and De Greef, appeared in 2016 and has been used in various courses at the IIS and elsewhere, which has provided us with a wealth of experience and feedback from both students and faculty members who have worked with it. For this second edition, therefore, we have extensively revised the text and several key elements of the interdisciplinary research process, which is the cornerstone of this handbook.

The first part has in fact been almost entirely rewritten by the handbook's first author. The chapter on interdisciplinary integration has been conceived anew – given that it is the key ingredient of interdisciplinary research – and is now more in line with the practice of interdisciplinary research than in the previous edition. It now presents a toolbox with a wide variety of integration techniques and explains why many research projects might apply a plurality of such techniques in parallel. This type of plurality is now addressed more explicitly in other chapters as well. For example, more attention is given to transdisciplinarity, as the inclusion of extra-academic stakeholders in research projects has become more prevalent and important in recent years. In addition, this edition makes explicit the multiple dimensions on which cases of interdisciplinary research can vary from each other. Although the second part of the handbook is less extensively revised, we have made some changes to the structure of the interdisciplinary research process to better reflect how this process works in practice, based on what we and our students and colleagues have observed. This edition also offers some insights on collaboration provided by members of an interdisciplinary research team, given that most interdisciplinary research is in fact carried out in teams. Finally, in contrast to the first edition which did not include references to projects involving the humanities, we have now included examples and observations from a wider range of disciplines and fields of research.

The first edition benefitted from feedback and commentary from a large number of (international) colleagues and students. Given the extensive revisions of this edition, we are grateful to the many people – from the IIS and elsewhere – who were this time willing to read our drafts and provide useful feedback and critical commentary. In alphabetical order, they are:

- Prof. L. Bertolini – IIS, Faculty of Science, UvA
- Ms. L. de Greef, MSc – IIS, Faculty of Science, UvA
- Dr. P.D. Hirsch – Department of Environmental Studies, SUNY College of Environmental Science and Forestry
- Dr. L. Jans – Environmental Psychology, Faculty of Behavioral and Social Sciences, Rijksuniversiteit Groningen
- Dr. E.F. Strange – Institute for Environmental Science (CML), Faculty of Science, Leiden University
- Prof. J. Thompson Klein – Emeritus professor of Humanities, Wayne State University
- Dr. J.C. Tromp – IIS, Faculty of Science, UvA

We would like to take this opportunity to thank the interdisciplinary researchers who shared their stories about interdisciplinary careers in Chapter 12:

- Dr. B. Vienni Baptista – Department of Environmental Systems Science, ETH Zürich
- Dr. A. Beukenhorst – Strategic Lead & Academic Liaison, Leyden Labs
- Prof. C. Lyall – School of Social and Political Science, University of Edinburgh
- Dr. J. Swart – Science, Technology, and Society, Michigan Institute of Technology

We are also grateful for the support and feedback of the lecturers and students of various bachelor's and master's programs, in particular:

- Ms. J. Libert, MSc – IIS, Faculty of Science, UvA
- Mr. A.N. van Woerden, MA – IIS, Faculty of Science, UvA
- Mr. M.H. Strømme, MSc – IIS, Faculty of Science, UvA
- Mr. I.P. Oostrom, MA – IIS, Faculty of Science, UvA
- Dr. E.D.A. van Duin – Parnassia Groep
- Mr. W. van der Hoeven, BSc – Alumnus of Beta-Gamma Studies
- Ms. L. van Eck – Alumna of Beta-Gamma Studies
- Ms. N. van de Water, BSc – Alumna of Beta-Gamma Studies
- Ms. N. Rijneveldshoek, BSc – Alumna of Beta-Gamma Studies
- Mr. N. van Stee, BSc – Alumnus of Future Planet Studies
- Ms. F. Mostert, BSc – Alumna of Future Planet Studies
- Ms. F. de Haan – Student of Beta-Gamma Studies
- Ms. N. Kasirga – Student of Future Planet Studies
- Ms. A. Mul – Student of Beta-Gamma Studies

Finally, we thank our colleagues for their coordination of the writing process and for their comments regarding the improvements from the first print:

- Dr. Y. Hartman – IIS, Faculty of Science, UvA
- Ms. J. Conemans, MA – IIS, Faculty of Science, UvA
- Ms. E. Krooshof, MSc – IIS, Faculty of Science, UvA
- Ms. L. Dokter, BSc – IIS, Faculty of Science, UvA

Preface

An Introduction to Interdisciplinary Research is a handbook on interdisciplinarity and a manual on how to conduct interdisciplinary research. Although several books have already been written about interdisciplinary research that have provided rich theoretical descriptions of and hands-on approaches to this topic, this handbook is a more condensed resource focusing on students in the social and natural sciences. The most relevant comparison to draw here is with Allen Repko's seminal *Interdisciplinary Research: Process and Theory* (now in its third edition, with Rick Szostak, 2017). Repko's book served as an important source of inspiration and information for us. Having used Repko's book for several years in our interdisciplinary research seminars, we felt the need for another book that would differ in several respects from Repko's valuable book. While Repko primarily addresses undergraduate students of the liberal arts and sciences in the United States and Canada, our book mainly focuses on (European) undergraduate and graduate students with more experience in disciplinary research. This is why our book primarily contains examples of research carried out in European interdisciplinary programs. Furthermore, our book has probably more to offer to students of the social and natural sciences. We have also included a thorough description of the concept of complexity, which we – and others – consider to be a main driving force behind interdisciplinarity. A related and not insignificant difference to Repko's manual concerns size: we explicitly aimed to produce a more condensed book that is practical, to the point, and clear.

The book is divided into three parts. The first part – *The Handbook* – presents a brief overview of interdisciplinarity and provides more conceptual insights into the origins of and reasons for interdisciplinary research, what its key features are, when it can be applied, and why it should be applied. This is all in preparation for the second part of the book – *The Manual* – which focuses on the step-by-step process of interdisciplinary research, setting out instructions on how to undertake this type of research. The third part contains a model example of an interdisciplinary project and a chapter highlighting the careers and experiences of some interdisciplinary scholars.

Many questions surround interdisciplinary research. How does it differ from disciplinary research? What does it demand of the interdisciplinary researcher? What possibilities does it have that disciplinary research does not offer? It is important to note that interdisciplinary research builds on disciplinary research. When dealing with complex problems, however, an approach that is merely disciplinary will not suffice. Such problems require an interdisciplinary approach to arrive at scientifically and socially robust answers.

The interdisciplinary research process is not an easy journey. In fact, it is a challenge for undergraduate/graduate students and experienced senior researchers alike. The aim of this book is to make the process more accessible. We provide many examples of interdisciplinary research projects, obstacles that researchers encountered during their academic journey, and the solutions they came up with. Moreover, we interviewed researchers who are experienced in applying an interdisciplinary approach, and we share their expert insights in this book.

As mentioned, it would have been impossible for us to write this book without the contributions of the experts, lecturers, students, and other individuals affiliated with the Institute for Interdisciplinary Studies (IIS) at the University of Amsterdam (UvA). We hope you learn much from reading this book and that you are able to put into practice any insights you obtain. We do welcome your feedback, so if you have any suggestions on how to improve this book (perhaps for a next edition), please get in touch with us at Onderwijslab-iis@uva.nl.

Chapter guide

The first part of the handbook begins with a short introduction that also explains why interdisciplinary research has been gaining in prominence (Chapter 1). We then briefly delve into the philosophy of science and offer a description of the science cycle, which is used later in the book to explain the nature of interdisciplinarity (Chapter 2). Chapter 3 continues with a philosophical and historical account of the emergence of disciplines and interdisciplinarity as well as a brief look at attempts at unification and pluralism. Pluralism is also covered in Chapter 4, which describes interdisciplinarity's variations and the drivers behind them. It includes sections on complex and wicked problems, transdisciplinarity, and action research. Chapter 5 introduces a toolbox of techniques for interdisciplinary integration – essential to interdisciplinary research – while following the structure of the science cycle that was presented in Chapter 2.

After reading Part 1, you will have acquired enough insight into and understanding of scientific research – interdisciplinary and otherwise – to start your own interdisciplinary research project. Part 2 will guide you through this process by means of a model of interdisciplinary research introduced in Chapter 6. The chapter points out where monodisciplinary and interdisciplinary research approaches differ and gives a step-by-step explanation of the process – from the definition of the problem (Chapter 7), the formulation of the research question (Chapter 8), and data collection and analysis (Chapter 9) to the discussion and conclusion (Chapter 10).

In Part 3, we provide an example of an interdisciplinary research project (Chapter 11) carried out by a team of students following the steps outlined by the model introduced in Part 2. Furthermore, we ask four interdisciplinary scholars to share their experiences with interdisciplinarity in Chapter 12.

Part 1
The Handbook
'The What'

1 Introduction

Half a century ago, philosopher of science Karl Popper famously observed: 'We are not students of some subject matter, but students of problems. And problems may cut right across the boundaries of any subject matter or discipline.' (Popper, 2002). Academic disciplines like anthropology, economics, history, mathematics, neuroscience, and physics are traditionally organized around the kinds of things that they investigate. Yet this division of disciplines assumes that we can understand or explain the properties of a specific 'kind of thing' or phenomenon from the perspective of a single discipline. As soon as we focus on a particular question or research problem that involves such things, however, we often find ourselves forced to collaborate across these traditional disciplinary boundaries.

Unsurprisingly, Popper's statement has become increasingly relevant. Today, many of the phenomena and problems that we are trying to understand and solve do indeed 'cut across' the traditional boundaries of academic disciplines. Whether we are focusing on phenomena as wide-ranging as cross-cultural communication, climate change, the financial crisis, genetic modification, an interpretation of a religious text, the Covid-19 pandemic, or life satisfaction, we will find scientists[*] from a wide range of disciplines working together to understand these phenomena and to develop responses to the challenges they pose. Such collaborations are a result not only of the growth of our knowledge, laying bare the connections between phenomena, but also of the growing complexity of our world, which creates more and more interdependencies. Both these developments – our growing knowledge as well as the increasing complexity of reality – compel us to give an ever-greater role to interdisciplinary approaches to research.

This growing importance of interdisciplinary knowledge was signaled by a groundbreaking 1972 report by the Organisation for Economic Co-operation and Development (OECD) called 'Interdisciplinarity: problems of teaching and

---

[*] Unfortunately, there is not a single word in English that refers to academic researchers in general, unlike in the Dutch language ('wetenschappers') or in German ('Wissenschaftler'). Although the word 'scientists' is generally understood to refer to those working in the exact and life sciences only, we will use it here in the more general sense of those engaged in some form of academic or scholarly research, including those in the arts and humanities and in the social sciences.

research at universities' (Apostel, Berger, Briggs, & Machaud, 1972). Since then, numerous interdisciplinary research and educational programs have emerged at universities and similar institutions across the globe. These developments are further fostered by academic institutions and funding agencies specifically aiming to support interdisciplinary research in a growing recognition of the added value of interdisciplinary research alongside disciplinary research. The European Commission – responsible for large international research funding programs – wrote in 2004, for example 'It is also seen in the fact that the academic world has an urgent need to adapt to the interdisciplinary character of the fields opened up by society's major problems, such as sustainable development, the new medical scourges and risk management.' However, the commission acknowledged that universities – and other organizations, we might add – find it difficult to adapt to this need for interdisciplinarity: 'Yet the activities of the universities, particularly when it comes to teaching, tend to remain organized within the traditional disciplinary framework' (European Commission, 2004, 11-12).

Given this prevalence of a disciplinary framework, how are we to understand the growing prominence of interdisciplinarity? As mentioned, it is not exceptional for a phenomenon to be determined by many different factors and for changes in the context to also have an impact, making it challenging to investigate the phenomenon and to seek to explain, predict, or intervene in it. Scientists often work hard to create a situation in which they can focus on one single factor – or only a few factors – contributing to the phenomenon. To this end, they have developed research methods that allow them to focus exclusively on one or several factors, for example in the laboratory where they can control the circumstances. Such focused research can lead to separate theories that describe – and possibly also explain – the relation between the phenomenon at stake and a single specific factor or a few such factors. However, and this is an important point, in our messy world, the same phenomenon might be affected by a multitude of factors, making it more difficult to investigate. If scientists succeed in accounting for all relevant determining factors as well as the interactions between these factors, they will be better able to understand, predict, explain, and perhaps even control that phenomenon.

Let's look briefly at an example to illustrate this. The link between alcohol and aggressive behavior has been known to mankind for a long time, as ancient texts and art works testify. However, more recent studies have made visible the causal pluralism involved in this connection, several of which are reviewed by Heinz and colleagues (Heinz, Beck, Meyer-Lindenberg, Sterzer, & Heinz, 2011). Instead of a monocausal link between the consumption of alcohol and aggression, the authors argue that a more comprehensive explanation involves multiple determining factors that also work in different forms. Some factors even play more than a single role in this connection. For example, there are various cognitive processes involved. As is well known, alcohol reduces the control that a subject is able to exert over his cognitive and behavioral processes, making him more liable to impulsive actions. Another cognitive effect of alcohol is a reduction in the subject's ability to steer his attention, which may result in a limited overview of a situation. Furthermore, alcohol impedes threat-related information processing, which can lead to wrong interpretations of another person's behavior. Finally, some individuals expect to become more aggressive upon alcohol consumption, which makes them act in a more hostile manner after being merely exposed to alcohol-related priming stimuli.

Although the analysis above focuses on cognitive processes and foregrounds relevant disciplines, other fields of research that Heinz et al. leave out could also have an effect. For example, the expectations about the effect of alcohol consumption on an individual are partly based on socio-cultural information and education. For example, the ancient Greek god Dionysus, who represented wine and theater as well as ecstasy, was also considered a liberator. However, as liberating as wine consumption and other forms of ecstasy might be, these often lead to violent and even tragic events as generations of spectators have learned from ancient and modern theatre plays. Looking at neurobiological factors, geneticists have shown that some individuals are more at risk of displaying aggression when drinking alcohol because their genotype specifically affects the functioning of their amygdala and hence emotion processing. Interdisciplinary research can further enrich our insights into the link between alcohol and aggression beyond those mentioned above. Although it may appear that we are unnecessarily complicating an already complex relationship, the hope is that adding such insights will help us explain additional variations in the aggression that some individuals demonstrate upon consuming alcohol. Moreover – and related to this – interventions developed to mitigate aggressive behavior in alcohol users must take such complexity into account if they are to provide a robust response that works not only under controlled clinical conditions but also in real life.

This example demonstrates that an interdisciplinary approach to alcohol-related aggression is necessary if we aim to develop explanations and predictions that are robust – i.e., that are valid under various conditions. We must be able to explain the relation between alcohol and aggression as it pertains to not just a very limited and specific group – for example, those who share a particular genotype and who are prone to specific cognitive responses – but also a broader group, and this means we must understand how the variations in response patterns are determined by multiple factors. Hence the need to invoke a plurality of theories and methods and to integrate multiple sets of data. If this proves insufficient, we could take a transdisciplinary approach. In transdisciplinary research, the net is widened to include not only scientists from different disciplines but also extra-academic stakeholders (Hirsch Hadorn et al., 2008). In this example, alcohol users and their family members and colleagues might be invited to participate in a project that aims to develop an intervention that is effective not just in controlled settings but also in real-world situations.

Designing a socially robust measure obliges us to consider the perceptions, priorities, and interests of all these stakeholders from the moment the initial research question is formulated up to when an adequate intervention is developed. Since the effectiveness of any intervention generally depends on the adequate compliance and collaboration of the target group and those in their environment, including stakeholders in the research project turns out to be crucial. If we look at Figure 1, we see how the perspective of an extra-academic stakeholder can help increase the robustness and relevance of scientific research. Although this undeniably puts additional demands and constraints on such projects, especially when performing such transdisciplinary research, we consider this transdisciplinarity sufficiently important to include it in this handbook on interdisciplinary research. For this reason, we will strive to prepare the users of this handbook for a wide variety of projects in which multiple boundaries are crossed – not just between disciplines but also between science and the world of lived experience.

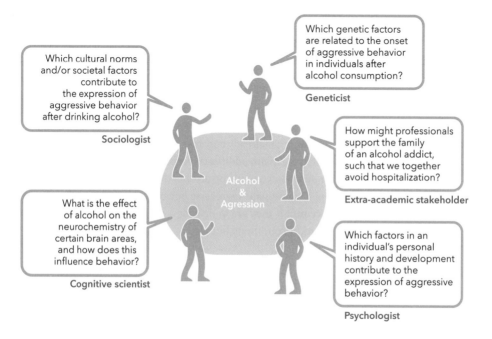

Figure 1   *Different perspectives on the relationship between alcohol intake and aggression*

Scientists conducting interdisciplinary research must have a theoretical and philosophical understanding of what a discipline is and how science in general operates. Such philosophical insights will help them to recognize and understand disciplinary differences and similarities and to grasp what integrating different disciplinary perspectives implies. Chapter 2 offers a tailored introduction to the philosophy of science and takes a close look at the science cycle, which most scientists employ implicitly or explicitly. The chapter includes a brief discussion of the theoretical and methodological pluralism that is common in science nowadays, which can facilitate interdisciplinary collaboration. It closes with an exposition of ontological, epistemological, methodological, and normative assumptions that are at stake in every research project, the articulation of which can help researchers remove obstacles and build bridges to interdisciplinary collaboration.

Chapter 3 then offers an account of the way in which disciplines have come to be categorized. This categorization is partly the result of historical and at times coincidental developments, which is why disciplinary boundaries might require revision – as Popper's quote implies. In line with Popper's student Thomas Kuhn, we argue that a discipline is not just characterized by its shared body of knowledge and methods but also by its social and institutional manifestations, including textbooks, conferences, and educational programs. These social and institutional structures engender both disciplinary specialization and isolation, which presents challenges to scientists investigating the complex topics mentioned above. Chapter 3 closes with a brief discussion of the movement to unify science – a failed attempt to overcome disciplinary specialization and isolation.

The concepts of multidisciplinarity, interdisciplinarity, and transdisciplinarity are defined in Chapter 4, which then goes on to delineate the several dimensions on which interdisciplinary projects can be distinguished from each other, such as: narrow or broad interdisciplinarity; the number and relevance of the disciplines involved; case-based or theory-driven interdisciplinarity; and their levels of integration. The chapter continues with an explanation of the increasing prevalence of interdisciplinarity and ends with a brief look at some recent developments affecting interdisciplinary research both in terms of content and methodology: complexity, wicked problems, transdisciplinary research, and action research. By showing these variants of interdisciplinarity as well as the latest developments in this field, we hope to help you reflect on the possible shape of your own interdisciplinary project and make informed decisions about this.

Finally, the more foundational part of this book (the 'what') closes with Chapter 5, which deals with the essential ingredient of interdisciplinarity: the integration of different disciplinary contributions. In line with the theoretical and methodological pluralisms discussed in Chapter 2, we present a toolbox of integration methods ranging from conceptual integration to the development of an interdisciplinary intervention or instrument. Importantly, a given interdisciplinary research project might employ multiple integration methods in parallel or during different stages of the project.

After having familiarized yourself in Part 1 with the structure and process of science and the ways in which disciplinary perspectives might be integrated with each other, you are now ready to incorporate these insights into your own project. Part 2, 'The Manual', presents our model of the interdisciplinary research process and its nine steps. Distinguishing the research trajectory into the orientation phase, the theoretical analysis phase, the data acquisition and analysis phase, and the completion phase, Chapters 6 to 10 guide you through the interdisciplinary research process – from adequately determining your research problem to interpreting your results and drawing conclusions. Each chapter presents several examples from the research literature to illustrate the steps to be taken and decisions to be made. The accompanying reflection questions help you to understand these steps and to make decisions together as a team. It is worth emphasizing here that, given that this is an 'iterative decision-making process' (Newell, 2007), during your interdisciplinary research you may need to revisit a previous step in light of the insights you have gained along the way. Indeed, interdisciplinary research is sometimes more time-consuming and frustrating than disciplinary research, but this should not surprise you considering its richness and complexity.

Since this book is based on the authors' shared expertise in supervising hundreds of undergraduate and graduate interdisciplinary teams, it also addresses the issue of team collaboration. Although it is not impossible for one individual researcher to integrate different disciplinary perspectives, disciplinary specialization and isolation make this the exception rather than the rule nowadays. Unfortunately, disciplinary training does not always prepare students sufficiently for working together with colleagues from other disciplines. Assuming that all are sufficiently curious about other perspectives and open-minded about crossing boundaries between disciplines, we will offer insights and practical advice on how best to work in a team. We are confident that, upon reading this handbook and applying its contents in practice, interdisciplinary research teams will be able to develop adequate solutions to the challenging problems Popper was referring to.

2 What is science?
A brief philosophy of science

2.1 What is science?

Alcohol consumption leads to more aggression: this seems to be a no-brainer. However, in the previous chapter we discovered that even such an apparently simple causal connection can give way to a more complex interaction of factors. The interaction between genes, cognitive processes, and behavior might not be as surprising as the insight that cultural information – about the 'liberating' role of alcohol – might influence someone's behavior by merely nourishing their expectations before they have even consumed any alcohol. When taking a closer scientific look at the simple link between alcohol and aggression, we see that it is mediated by a host of extremely heterogeneous factors like genetic disposition, cognitive processes, forms of behavior, social relations, environmental factors, the interpretation of cultural information, and the many interactions between these factors. Investigating all these factors scientifically requires that we consider a broad variety of relevant theories and concepts, employ extremely heterogeneous methods, and interpret a wide range of results regarding these factors' contribution to explaining this link between alcohol and aggression. How can we make sense of all of this?

For navigating and integrating such a divergent set of theories, methods, and insights, it is extremely useful to make use of the conceptual 'toolbox' that philosophy of science provides. As scholars engaging in a 'second-order activity', philosophers of science reflect on the first-order activity that scientists from theologians to urban planners undertake. A philosophical analysis allows us to examine their activities at a more abstract level and, for example, make explicit the assumptions that many scientists take for granted while doing their job – assumptions about the correct research methods, about whether quantitative or qualitative data are adequate, about the real-world applicability of scientific insights, and so on. After articulating such assumptions, we are also able to consider the similarities and differences between scientific disciplines. The ability to see connections is necessary if we intend to bring several disciplines together in our work. For example, each discipline implicitly foregrounds some factors it deems relevant and presents corresponding methods to investigate these factors, while other factors are relegated to the background and treated as contingent factors. Cognitive neuroscientists will assume that behavior is always dependent on cognitive processes and might prefer to investigate these using a combination of psychological questionnaires and brain imaging experiments. Yet their colleagues in sociology and anthropology will assume instead that power relations and cultural norms largely determine the contents and outcomes of these cognitive processes, leading them to relegate the cognitive processing itself to the background.

That different scientists might contribute to our understanding of a particular problem even though they approach it from perspectives that are seemingly at odds with each other is what made philosopher of science Karl Popper observe a certain tension between the way in which scientific disciplines are organized and how problems present themselves to us. Indeed, interdisciplinary research is one way of overcoming this tension and of reorganizing scientific research so that it is not impeded by the existing organization of the sciences. Since it is important to understand both the value and the limitations of this organizational structure, we need to briefly reflect on what science is and does. In other words, let us reflect on some of the basic ingredients of science in the way that philosophers of science do. There are many ingredients that appear to be familiar enough, though perhaps not easy to understand, such as theory, concept, fact, hypothesis, explanation, inference, induction, deduction, and so on. Given the limited scope of this handbook, we examine only a few of these ingredients and therefore recommend that you look elsewhere if you seek a more comprehensive introduction to the philosophy of science.

First, it is important to acknowledge that scientists, like all humans, generally rely on their sense perception to acquire information about the world. They will subsequently consider this information carefully to draw conclusions while avoiding flawed arguments or logical mistakes. Nonetheless, scientists do this in quite peculiar ways, as revealed in the texts they produce. Scientists tend not to rely only on their senses but instead use a variety of fancy instruments to perceive more, smaller, larger, and different objects than lay persons can, using microscopes, structured interviews, satellites, fMRI scanners, validated questionnaires, participatory observations, archival research, cyclotrons, big databases, and so on. Similarly, their reasoning and arguments are specialized, working as they do with specific concepts, propositions, mathematical formulas, figures, tables, electro-circuitry schemes, and so on. In addition to these special devices, their reasoning is also uncommon, as it is guided in specific ways by the rules of logic. In other words, although scientists gather insights in uncommon ways, and even if their arguments contain rather specific contents and concepts and they employ stricter rules of logic than lay people usually do, their aim is merely to reason and communicate as specifically, clearly, and correctly as possible.

Related to this specialized way of gathering and reasoning about information is the fact that scientists are expected to make explicit how their work is building upon the work of their colleagues. In turn, they must subject their own newly developed insights to the collective scrutiny of their colleagues – by publishing it in peer-reviewed journals and presenting it at academic conferences. As one of the most famous scientists once said: 'If I have seen further, it is by standing on the shoulders of giants.' (Isaac Newton). Even though every scientist works hard to develop a novel insight, part of that work involves making explicit the current state of relevant knowledge pertaining to the problem at stake, expanding the knowledge base that others have produced, using that knowledge in new applications, or proving that their colleagues' conclusions are incorrect and that adjustments are required, and so on. Yet where does one start when the aim is to establish the current state of knowledge about a problem? And how do we proceed after we have done that and want to contribute to that state of knowledge by filling an existing gap of knowledge that we detected? More pertinent still, for our purposes: how does one accomplish all this when we are tackling that problem or question from multiple perspectives, detecting various gaps in knowledge, and employing more than just one single instrument? Returning to the alcohol and aggression example: how do cognitive neuroscientists proceed when they want to explain the impact of a specific gene on a subject's alcohol-induced behavior, or classicists when interpreting from ancient texts the ancient Greek attitude towards such behavior?

In this chapter, we will present the so-called science cycle (also sometimes referred to as the empirical cycle) consisting of a series of consecutive steps that help us to illustrate the work that scientists are engaged in. Each of these steps consists of ingredients and procedures that together make up the scientific 'cooking process'. In a certain sense, the scientific research process is never complete, as consuming the food will never satisfy our hunger. Unlike in cooking, however, scientists are generally not making the same dish repeatedly, as they always change some of the ingredients, aiming to improve on previous work or complement it in some way. The word 'cycle' tends to obscure this fact, so that it might be more appropriate to speak of a 'science spiral' in which each turn brings us a step closer to the truth – 'approximating the truth', in the words of Popper – although new questions usually emerge during the process.

Understanding this 'science cycle' and its components not only better prepares scientists to articulate and reflect on the research within their discipline; it also enables the kinds of joint articulation and reflection that is needed whenever scientists join forces with someone from another discipline. Interdisciplinary collaboration requires some way of integrating the two scientific 'cooking processes', which can have consequences for more than just a single component. Explaining the components and procedures that make up your research, a team member from another discipline can consider where and how they might contribute to the process by adding or changing one of its components or adjusting a procedure. The figure below represents this process: one researcher articulates the components (and how they are related to each other) that together make up her research process, allowing her colleague to consider how he might contribute to that process – which in this case amounts to adding a component to it. This is a relatively simple form of interdisciplinary integration, but there are of course many other forms and many more complex forms. Chapter 5 is devoted to a systematic treatment of different forms of integration. Many interdisciplinary projects will involve multiple forms of integration in parallel, which implies that in an interdisciplinary team, the joint reflection and conversation on your research process is an ongoing task that should never be considered completed. We will see below how all the components and procedures included in the science cycle can be part of that reflection and conversation.

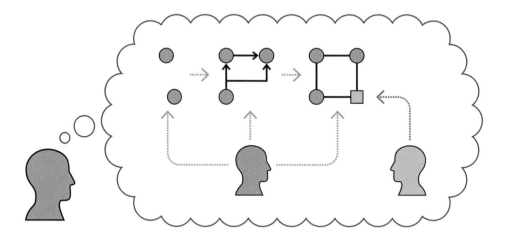

Figure 2   An expert engages in metacognition and philosophical reflection about his or her thinking. The expert reflects specifically on a learning process (in the cloud) and the set of representations it has yielded. Such second-order reflections also prepare the expert for the integration of an additional insight from another expert incorporated into the complex mental representations, such as the green square added here (Taken from Keestra, 2017). Reprinted with permission from the author.

2.2   Moving through the science cycle

Theories and laws

The beginning of the science cycle (the top of Figure 3) consists of the most significant component of science: theories and laws. A dictionary of science defines 'theory' as 'a set of ideas, concepts, principles, or methods used to explain a wide set of observed facts. Among the major theories of science are relativity, quantum theory, evolution, and plate tectonics.' (Lafferty & Row 1997). This definition is very short and specifically represents theories from the sciences, and as such, it captures an important function of theory, which is to bring together a previously non-unified set of observations in such a way that we come to understand how and why these observations hang together. Let us explore this somewhat further while keeping in mind that theories may bear 'family resemblances' across the humanities, the life sciences, or other disciplines. Hence, we must take a somewhat more liberal approach.

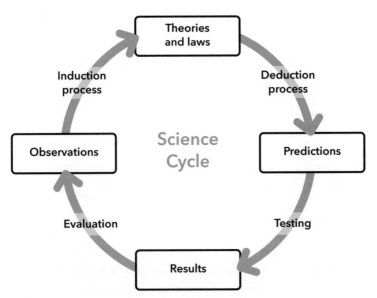

Figure 3   The science cycle, consisting of four processes, connecting four components – together providing a (somewhat simplified) representation of science as an ongoing process. Note that reasoning and sense perception are present, albeit in specific ways.

As the short definition states, scientists typically develop a theory after having collected a 'wide set of observed facts'. These facts could be observations of fear responses in rats navigating Morris' water maze, or interpretations of the word 'good' as it appears in many places in Homer's *Odyssey*, or descriptions of social interactions during traffic hours, or trails of neutrinos presumably detected in a reactor experiment. Although in most cases, scientists collect such observations after having read and used relevant theories previously presented by other scientists about these phenomena, at some point, induction also plays a role. Induction is a process of reasoning that starts from many statements – each of which covers a particular example or observation – and moves to a general statement. A famous example is the general statement that 'all swans are white' which might be presented by an ornithologist after having observed hundreds of individual swans, all of which were found to be white. This example immediately shows how risky the process of induction is, since no ornithologist can ever observe all swans everywhere and at every moment in time. Indeed, he may have overlooked the black swans that live on another continent. The observation of a single black swan would render the statement 'all swans are white' logically wrong. However, a theory does more than just bring together a set of observed facts into a single, general statement. It is the additional element of explanation, understanding, and/or prediction that differentiates a scientific theory from a common general statement, even though induction inevitably does play a role in both. This element is also what sets the science cycle in motion, forcing us to continuously improve, expand, and adjust the general statements that characterize our sciences.

Let us now consider a somewhat broader account of what a theory is and does. A theory consists of concepts, principles, ideas, or statements that together provide a comprehensive background or framework that helps to further determine the ingredients of science, like hypotheses that can be investigated and the methods to do so. The primary function of a theory is to explain, understand, and possibly predict a diverse set of phenomena by unifying these phenomena with the help of a certain pattern or set of relations between relevant elements. Familiar examples of theories include the behaviorist theory of learning, the structuralist theory of meaning, and the game theory of strategic action. Each of these theories was proposed after scientists had gathered many observations, interpretations, or other sets of data.

What is important to realize is that a theory should provide us with a better understanding of real phenomena or events that we can observe or test. Yet it is not uncommon for a theory to include one component or another that is itself still difficult to understand or explain. Gravity, for instance, plays a role in both Newtonian mechanics and Einstein's theory of relativity. In both theories, gravity entails a certain relation between masses and/or energy; however, the two theories understand this relation quite differently. Newtonian gravity refers to the attraction that two bodies exert upon each other, whereas Einstein's definition of gravity refers to space-time curvatures. In both cases, theoretical components enable us to understand, explain, and predict the movement of bodies in the universe, though each employs a different scope.

Although it might sound counterintuitive, some theories and laws in quantum physics are relatively simple – notwithstanding the strange theoretical entities and relations they propose – compared to the theories developed to capture complex and dynamic phenomena in the life sciences and the human sciences. Obviously, the four forces in quantum physics together produce astonishing phenomena and require complex mathematical descriptions. Yet they are simple in comparison to phenomena like evolution, consciousness, the emergence of religious rituals, and globalization because the latter are typically affected by numerous factors at the micro, meso, and macro levels, are often spectacularly multicausal and emerge from complex historical trajectories. Religious rituals, for example, can involve social groups like citizens and priests, can follow culturally specific cycles of life, might employ natural resources and hallucinogenic substances, are responsive to historical events, and address psychological and social emotions and relations (Burkert, 1985). Each of those factors might invite a separate theoretical treatment, enabling us to understand and perhaps predict at least partially why some rituals emerge, how they are experienced and performed, what causes them to transform, and so on.

Clearly, the more comprehensive a theoretical framework is, the more we must consider such factors and their potential interactions as well as their responsiveness to pertinent environmental and historical conditions. This is why interdisciplinarity plays an important role in such contexts. If we need to integrate elements and relations that are generally studied in separate disciplines, their integration into a theoretical framework requires careful attention: do the contents of theories from different disciplines translate easily into each other? Do we understand their interactions and relations? Or is it impossible for the elements and relations covered by one discipline to be connected to or identified in some way by elements of another? Do cognitive neuroscientists define levels of consciousness in the same way as scholars of shamanic rituals do? We will discuss the challenges of such integration more closely in Chapter 5, but for now it is sufficient to be aware of the conceptual challenges involved in creating an interdisciplinary theoretical framework.

Finally, it is important to realize that there are differences between disciplines in terms of the prevalence and importance of specific theories and laws. In some disciplines, researchers agree that a few theories cover most of their research domains, while in other disciplines, many theories are available. Physics, for example, is dominated by the theories covering classical and statistical mechanics, special relativity, quantum mechanics, thermodynamics, and electromagnetism. Although the theories have some overlap with each other and are in some ways not completely consistent with each other, they are used next to each other, as they largely complement each other. Importantly, most physicists will agree about the status of these theories.

By contrast, in the social sciences and humanities, we find many competing and sometimes even contradicting theories. Behaviorism, for example, is a theory that contends that animals and humans learn from repeated stimulus-response chains or conditioning. Some scholars use this theory to explain human learning and behavior more generally. This puts them at odds with scholars who adhere to an existentialist theory, which maintains that humans are responsible for shaping their own 'existence' by deciding their preferences and actions. Covering largely the same set of phenomena, these two theories offer almost opposite explanations and present different elements in those explanations: learning, stimulus, response in the former, and options, responsibility, decision in the latter. As much as two such theories appear to contradict and exclude each other, there are often ways to adjust and integrate such theories in these domains. For example, cognitive behavioral therapy helps patients to recognize unhealthy patterns of conditioned behavior and thoughts while assisting them in imagining new patterns of thought or behavior they might decide to adopt in order to overcome phobias or depressing ruminations.

It is important to realize that theories in the life sciences, the social sciences and humanities often foreground a particular property or factor which can easily coexist with many other properties or factors. Human behavior, for example, can be influenced at some point by money, by jealousy, by aggression, by love – or by a complex mix of such factors. Indeed, given the multifactorial or multi-causal nature of social, cultural, and historical phenomena, these allow a wealth of theories, each of which may explain a portion of the complex and dynamic nature of such phenomena. Hence the abundance of theories – and 'isms' – in those domains and the relative lack of precise laws.

In many cases – though not in all – we consider laws as part of a theory: for example, the law of gravity and several laws of motion belong to the theory of classical mechanics formulated by Newton. Somewhat differently, the Mendelian laws of inheritance are now part of the theory of classical genetics. Laws generally specify the relations between the elements that play a role in these theories. Note that the Mendelian laws of inheritance have a probabilistic character, which distinguishes them from laws belonging to classical mechanics and many other laws. Indeed, in the life sciences and the human sciences, such probabilistic laws are more prevalent because of the more multi-causal nature of phenomena in these domains. This also underlines the importance of distinguishing between correlative relations and causal relations: if we do not really know what causal mechanisms are involved, we should be cautious about interpreting probabilistic relations.

Deduction leads to novel predictions

As mentioned earlier, besides sense perception, reasoning is crucial to the scientific process, with induction and deduction being the most prominent reasoning processes involved in science. Induction helps us to jump-start the science cycle, as it enables us to move from observations of many individual cases to a statement about general patterns that cover these cases – a statement we hope to demonstrate to be valid. Typically, though, a scientist who wants to add something new to a body of knowledge will start by deriving novel predictions from an existing, already tested theory – or from a combination of two different theories. She will then carefully try to formulate predictions via a particular logical reasoning process called deduction.

Deduction enables us to derive specific statements from a general statement or theory, making it the reverse of induction. A simple example is the following. The theory of classical mechanics implies that bodies with mass exert a force on each other. From this, it can be logically deduced that the earth exerts a force on the moon, that a ball will drop to the floor, etc. Given that a theory covers a set of elements with properties – specified according to the theory – and describes one or more relevant relations between those elements, deduction implies applying the more abstract theory to a specific case. If correctly applied, no new information is added during this process. Classical mechanics predicts that the earth and moon attract each other because of their masses. The logic of deduction precludes us from deriving from the theory of mechanics a statement like 'all humans are mortal' because the elements and relations involved in mechanics do not contain humans (perhaps only humans as 'having' bodies with mass) nor their lifespan. However, in many instances, the validity of deduced statements is not as easy to determine. Given the important role of deduction in the science cycle, we must use it cautiously.

Deduction can lead to a new insight when previously unconnected statements or theories are combined with each other, even though no new information is added to this mix. In science, the next step would then be to test this new insight. If we look again at the case of alcohol and aggression, for example, an interdisciplinary scientist may deduce a specific prediction when combining a cognitive neuroscientific theory with a neurochemical theory and a theory from the science of religion. This novel combination of theories might lead him to deduce several new detailed predictions, for example about how alcohol use might lead to more aggressive behavior which could be observable during ritual sacrifice, lending itself to multiple lines of experimental, clinical, social scientific, and theological research. When applied to the slaughter of animals in some religious rites, she may indeed find that the sacrifice is performed differently after alcohol consumption than in alcohol-free cases.

In the previous section, we warned that induction carries a risk, as it jumps from a limited set of observations to wholesale statements like 'all swans are white'. Deduction does not per se involve such a risk. However, it can be used incorrectly. When scientists make incorrect deductions, this creates serious flaws in their subsequent research. A simple case of an incorrect deduction is when the statements 'all swans are mortal' and 'Socrates is mortal' lead someone to conclude that 'Socrates is a swan'. Even though both statements are correct, and birds and humans do share the property of being mortal, it is wrong to derive from this that Socrates and swans are identical.

It is not always easy to detect how flawed deduction can impede scientific research. In the case of the influence of alcohol on religious ritual, for example, the prediction that alcohol-free sacrifice is performed less aggressively may not be supported by evidence from a series of observations. Although the deduction itself might be correct, the statements involved in the deduction might need to be revised, as they are perhaps not all correct. Indeed, upon closer scrutiny it may turn out that aggression is not equally involved in all religious sacrifices, as sacrifice in some religions entails a form of elevation of the animal involved – almost as a form of care. So, if one of the theories involved – e.g., 'all sacrifice entails an act of aggression' – is not correctly understood or even mistaken, the deduction will also be flawed, and the results of the ensuing research cannot be valid.

It is relevant to note here that the process of deduction in the life sciences, the social sciences, and the humanities often yields results that are different from other scientific domains. This is due to the fact that phenomena in these domains tend to be determined by a host of very different factors, providing room for many theories – and 'isms' like structuralism, behaviorism, and Marxism – each of which may have some bearing on these phenomena. In these domains, it is similarly 'easy' to deduce a novel prediction, which can then be investigated or tested. Yet to determine the weight of the tested factor is not easy at all, given the fact that many different factors may be operating in parallel. Indeed, it may also be difficult to take one particular factor and test or scrutinize it independently of other factors. Returning to the example of animal sacrifice, the aggression involved in a ritual might depend not only on prescribed alcohol consumption but also on the historical origin or symbolic meaning of the ritual or its conjunction with specific social tensions. If the sacrificial animal represents an evil being, it will be treated differently than if it represents a religious idol or a powerful king. Thus, depending on the theoretical account of the animal, scientists may deduce very different predictions about the meaning and consequences of its sacrifice.

Nonetheless, the distinction between complicated cases like this and those in the natural sciences should not be overestimated. In both realms, we might discover that more factors contribute to the phenomenon than we initially presumed. For example, if we are going to test the predicted force of attraction between two bodies and are not aware of the fact that electromagnetic force might counteract the gravitational force, we would be equally misled in our oversimplified deduction. As correct as our deductive reasoning might have been, the statements on which it is based might misrepresent or oversimplify the phenomenon and therefore be flawed.

Testing to obtain results

Assuming that our deduction leads in a correct and adequate manner to a novel prediction, according to the science cycle it is now time to test this in the real world. However, before a scientist can put her prediction to the test, she needs to think creatively about a reliable way to go about this. It is one thing to deduce from the existing body of knowledge that there must be a connection between x and y; it is quite another to develop a way to investigate this in a specific way. In the case of our example on the relation between alcohol and aggression, what behavior should be tested, and by whom? What kind of alcohol consumption should be used in the investigation? The scientist must think of how to *operationalize* her prediction and what type of investigation might be adequate to study this operationalized prediction. Such an investigation may involve conducting an experiment, but it could also require other methods and data such as archival research in criminal records, in-depth interviews with subjects, or participatory observation in night clubs. In order for an operationalization to be scientific, the scientist should aim for it to be something that can be repeated and reproduced by others in different locations and at other times.

There are many operationalizations that a scientist might choose, and various considerations might determine this choice. For example, given neurochemical pathways, what important neurochemical compounds associated with alcohol use and aggression can we investigate that are the easiest to track? What brain areas might be involved, and what instruments enable us to measure correlations between their activation patterns and the behavioral changes in our subjects? Who might we use as guinea pigs for such experiments or observations, and what is the appropriate amount (ethically and health-wise) of alcohol that they should drink, over what period of time? Looking ahead to the results, what operationalization yields the kinds of data that we can analyze given our statistical expertise and available instruments? Similarly, a humanities scholar or social scientist who has developed a prediction about the meaning of a particular sacrificial ritual, for example, or the power relations that influenced its symbolism, should pose the same kinds of questions. The humanities scholar might want to operationalize her question by focusing on its appearance in certain paintings, requiring her to combine insights from iconology with insights about pertinent texts to suggest a probable interpretation of the ritual at stake. The social scientist, in turn, might want to focus on a particular religious crisis and seek to understand it by employing a structuralist theory of power. In other words, scholars of the humanities and the social sciences also need to determine how their deductions from general theories might be turned into concrete objects of investigation.

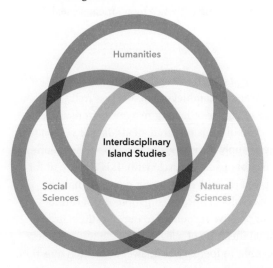

Figure 4 Venn diagram of the position of interdisciplinary island studies at the interface of social sciences, natural sciences, and humanities (Taken from Norder & Rijsdijk, 2016).

What this brief reflection on operationalization demonstrates is that there are usually many ways possible to investigate or test scientific predictions (explained in detail in Chapter 9). Especially when we are conducting interdisciplinary research, it may make sense to use multiple operationalizations, for it allows scientists to scrutinize their predictions in different ways, each of which is methodologically independent from the others: archival research, for example, uses different sources of information and focuses on different kinds of data than experimental lab research with subjects or participatory research in night clubs. Multiple operationalizations are effectively a form of *triangulation*, in which a phenomenon is pinpointed from three or more angles. In interdisciplinary research, these angles would be the different disciplinary perspectives, each of which requires a different operationalization. The figure shown above depicts three disciplinary perspectives that are used in interdisciplinary island studies. Investigating the tourist meccas or ecological refugia that islands may represent requires operationalizations from at least these three perspectives. Clearly, for a specific study, scientists from the three disciplines must jointly formulate a specific testable prediction that contains elements from and is consistent with these theories. For example, in some countries people have certain associations with the island of Mauritius, which may explain why it has attracted many visitors, which in turn has contributed over time to the destruction of much of its original forests (Norder & Rijsdijk, 2016). Note that in formulating this prediction, the scientists are also performing a particular form of interdisciplinary integration – a subject that is dealt with in Chapter 5.

Having first deduced a specific hypothesis or prediction from our theory and subsequently developed one or more operationalizations, we can now test these and check whether or not the results of our investigations confirm our original prediction. Yet the data obtained from our testing activities by no means speak for themselves, as we will see.

Evaluation turns results into observations

Having deduced from one or more theories a particular prediction and having operationalized this prediction, we proceed to conduct our actual research: investigating a particular object, for example, or collecting responses from subjects, or developing the interpretation of a symbol. This research yields data or results that first need to be evaluated before we can apply the emerging insights to the theories and laws that we started out our empirical or science cycle with. It is important for such an evaluation to be conducted properly, with sufficiently rigorous analysis and interpretation, without which the results of our investigation or test cannot provide us any insight. The number of positive responses to our questionnaire, the temperature of a patient, a potential meaning of a legal term, the tests of a new compound in the lab: these results cannot in themselves answer a research question, as they require more context for the assessment of their validity and significance.

So, what do we need to look out for when evaluating the results of a scientific investigation? As with everyday observations, when we are conducting a scientific investigation, our sense perception is inevitably involved in some way or another. This can occur via direct sensory observation of carefully orchestrated experimental events or of specifically selected human interactions, or it can happen indirectly when special instruments like neutron detectors or fMRI scanners are developed to detect or measure specific processes or objects. When perception itself plays a role in science, the operationalizations and instruments involved implicitly build upon the theorizing, deductive reasoning, and operationalizations that other scientists have engaged in – sometimes generations before us – creating those instruments and tests. Indeed, philosophers of science refer to the way in which scientific observation becomes entangled with the other processes of the science cycle by referring to the 'theory-ladenness' of observations: our observational methods and instruments often implicitly assume our acceptance of other theories and operationalization methods.

Something similar holds for the evaluation of the results or data obtained from our research. In order not to take these results at face value, scientists have developed sophisticated methods for their analysis. Depending on the nature of the prediction and the method used to generate our results, we must choose the most adequate methods of analysis. Such a method could involve a statistical method to detect significant patterns in our experimental data, the use of a Munsell color coding scheme to determine the nature of the patches of soil taken in a geological study, or a hermeneutic method for converging on an interpretation of a ritual that figures in several texts and images we had selected. Based on the careful and methodological evaluation of our results, we might conclude that we have discovered a certain 'fact' – fact taken in a more technical sense. Again, our methods of evaluation presuppose insights and results from scientific work carried out before us and tacitly accepted by scientists in the field.

It may now be apparent why it is mistaken to believe that there are 'naked facts' or data that in and of themselves will lead any sensible person to a particular conclusion. Instead, we should recognize that even unscientific or lay perceptions are dependent on the knowledge and beliefs that people already had before their observations. Most adults will agree, for example, that the perception that the Earth is flat is an illusion caused by the fact that it is nearly impossible to perceive, with the naked eye, the slight curvature of the earth – although children might still insist on the earth being flat. While most adults know about the curvature of the earth when perceiving the horizon, there are many ways in which adults – scientist or not – are still subject to misinterpreting what they perceive. Obviously, such misperceptions or misinterpretations can have a seriously negative impact on our scientific insights. A striking and sobering example is the under-diagnostication and misinterpretation of heart failure in women because the medical field did not sufficiently recognize the impact of hormonal changes or gender differences on the symptoms involved (Stramba-Badiale et al., 2006). By now, most cardiologists acknowledge that these physiological factors explain the observed differences in symptoms and apparent prevalence of heart failures instead of factors they previously attributed these differences to, such as gender-specific psychosomatic factors or sensitivity to pain.

It is only by properly evaluating the results of our investigations that these results turn into genuine scientific observations that play an important role in the science cycle. Naturally, our observations respond in some way to the predictions that emerged at an earlier stage in the cycle, which we set out to test: they confirm or disprove the predictions. Given the complexity of the steps leading up to our observations – i.e., the operationalization and evaluation steps – there are various reactions possible whenever our expectations or predictions are not confirmed by those observations. We might reconsider our intermediate steps to check for mistakes or flaws, for example by checking the adequacy of our instruments or the correctness of our calculations. If such reconsiderations do not lead to a rejection of our observations, a final step awaits us, bringing us back again to the top of the science cycle: to the theories and laws that characterize science.

Full circle: observations modify theories
Regarding the component at the top of the science cycle – theories and laws – we mentioned earlier how inductive reasoning initially led to their formulation. We also noted that induction is a process that entails some risks, given that we cannot test all potential occurrences of an event under all potential conditions (e.g., perceive all swans – white or otherwise – living now and in the past somewhere on the

planet). Due to the nature of inductive reasoning, any resulting theory bears some inherent 'risk' or uncertainty that our later investigations might bring to the fore. Thus, when we first test a prediction that we deduced from our theory and find out that our carefully evaluated observations do not concur with our initial theory, these observations might compel us to reject or adjust our theory.

Indeed, Karl Popper encouraged scientists to formulate their testable predictions in such a way that they would lead to the rejection of a theory. Considering the precarious nature of induction, which implies that we can never infer a general statement with utter certainty from a collection of observations, Popper concluded that scientists should give up trying to prove a theory's truth. Instead, he urged scientists to try to disprove theories, an approach that makes more logical sense. According to Popper: scientists should devote their efforts to refuting or disconfirming a theory instead of verifying or confirming it, as this would help to weed out incorrect theories. It only takes a single observation – of a black swan, for example – to refute a theory. That single observation would be sufficient to teach us an important lesson: that our current theory – that 'all swans are white' – is inadequate. Notwithstanding the logic behind Popper's recommendation, scientists do not adhere entirely to such an approach, preferring instead to adopt two strategies – i.e., checking whether a particular theory can be easily refuted or disproved while at the same time gathering new results that would support that theory. Indeed, instead of rejecting a theory if the results appear to refute it, scientists tend to check whether such negative results are simply the result of failed instruments, flawed analyses of the results, an unrepresentative group of subjects, a skewed selection of texts that have been analyzed, and so on.

Generally, though, scientists' observations will not amount to a simple confirmation or rejection of a certain theory. Rather, observations obtained in new experiments allow scientists to expand a theory's domain somewhat, or to adjust it to new conditions, or to add some further details, or to otherwise fill some gap in our knowledge identified earlier. Scientists may, for example, discover that the theory covers a wider domain than initially thought and that it applies to more groups of subjects or under a wider range of conditions or to other historical periods than were tested previously. Naturally, the converse is also possible: they may be compelled to narrow the scope of the theory and can now articulate its limiting conditions as indicated by their observations. Some theories may not apply to the whole population – as cardiologists found out only recently regarding women – or it may only lead to valid results when tested under certain conditions.

Interdisciplinary research offers scientists more options when testing a theory. As mentioned above, interdisciplinary research entails a kind of triangulation by applying different disciplinary approaches, each requiring its own operationalization and research methods. When such a technique as triangulation has been used and the observations yielded by the approaches subsequently all prove the existence and properties of a common object or phenomenon, this observation can be called *robust*. Robustness requires a 'variety of independent derivation, identification, or measurement processes', each leading to comparable and consistent results (Wimsatt, 2007). When geneticists, neurobiologists, and psychologists produce consistent observations concerning the relation between alcohol and aggression in a particular group of subjects even when using very different methods, their results mutually reinforce each other and lead to increased robustness. They may find that a particular group responds in a specific way to alcohol abuse, with that group bearing a particular set of genes that influences how alcohol is decomposed.

Interdisciplinary research like this enhances the robustness of scientific observations of this group's specific behavior during alcohol consumption by showing that this group is not only specific in its behavioral response but also shares specific genetic properties. Such research also allows scientists to integrate different theories into a more complex interdisciplinary one, thereby merging theories from different fields into a more comprehensive account of when and why alcohol can induce aggression. The observations made in this manner might thus enable scientists to induce that a certain group has a specific reaction to alcohol consumption because that group's genome carries several genes that together delay the physiological decomposition of alcohol, which implies that alcohol has a greater impact on that group's relevant cognitive processes. Clearly, this kind of inductive reasoning is much more complex than jumping to the conclusion that 'all swans are white', yet at the same time it is the result of many tests of separate theories by several scientists from different disciplines. In other words, bringing together results from different fields is quite different from simply collecting many white swans.

Some final remarks: pluralism and assumptions

Popper's critique of verification points to an important insight: merely repeating the same experiment or investigation continually might yield a huge collection of data and yet still fail to prove a theory's truth. Conducting a more robust scientific investigation requires the use of a plurality of methods and theories. As mentioned above, the complex and dynamic nature of many phenomena and the ambiguity of human expressions make it clear that such phenomena and expressions are determined by multiple interacting factors and sensitive to changing contexts. This suggests that, in most cases, a single theory will not be sufficient to explain and understand such phenomena exhaustively. In the case of the relation between alcohol and aggression, we must combine genetic, social scientific, and psychological theories. By the same token, our methods of investigation should be diverse. This means that if the connection between alcohol use and aggressive responses in our group of subjects is founded on investigations focusing not only on the psychological level but also on the level of the group and of genetics, our inference – or induction – to a causal connection is more trustworthy than if it rested upon a single research method and its associated theories. By its very nature, interdisciplinary research implies theoretical and methodological pluralism. This brings its own challenges but usually also yields more robust results than monism does.

Moreover, while triangulation adds robustness to the relevant insights, it is important to recognize that this is often not just a matter of quantity or of adding more supporting observations. In many cases, the addition of support from another line of research also implies a qualitative difference regarding our insight into the phenomenon. Compare the relatively simple phenomenon of stereovision, which involves combining two images from two separate observation points (or eyes) placed somewhat apart. This anatomical fact does more than merely add robustness to our ability to see, it also enables us to perceive something qualitatively new: the additional dimension of depth (see Figure 5 below). Similarly, interdisciplinary triangulation tends to do more than simply add robustness to a particular insight – it leads to new insights as well.

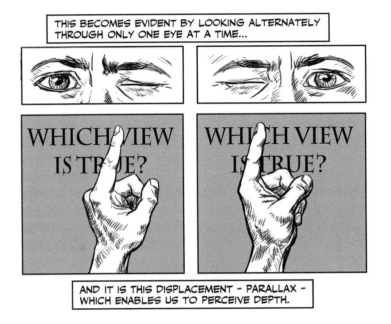

Figure 5   The integration of two slightly different perspectives on an object adds an additional dimension of information to the perceived images - which is analogous to interdisciplinary research offering more than multiple unrelated perspectives (taken from Sousanis, 2015). Reprinted with permission from the author.

The pluralism that we are advocating here and assuming throughout this book applies to the theories, the methods, and the nature of the results that scientists develop. Since phenomena like alcohol-induced aggression or ritualized behaviors are determined by a wide range of factors, there are a variety of approaches that can be used to investigate, explain, understand, predict, and modify such phenomena. Instead of seeing their own approach to a particular research field as the only logical one or assuming that their approach should prevail, researchers should appreciate the presence of such pluralisms. In the case of the study of consciousness, for example, neurophilosopher Patricia Churchland rejects the notion that the pluralism pertaining to this complex phenomenon betrays a weakness in the field of consciousness studies. Instead, she appreciates the presence of multiple theories because they offer a valuable multitude of 'lines of attack' to consciousness (Churchland, 2005). This multitude does, however, require scientists to spend more time and attention on articulating their approaches to colleagues from other disciplines than they would need to do when working within their discipline only.

In this context, it would be remiss in this chapter on philosophy of science not to delve into the topic of assumptions because they can hinder or facilitate interdisciplinary collaborations. Indeed, the pluralism discussed here is partly the result of assumptions. Like all people, scientists cannot but uphold many assumptions or presuppositions, as otherwise it would make science – and indeed life – almost impossible. In everyday life, as in science, we generally do not reflect on all the assumptions we make throughout our day, nor do we discuss them with each other. For example, when two people plan to meet, they do not discuss their assumption that the world will continue to exist, and that time will pass in a single direction and at an equal speed for both of them. Thus, even for as mundane a task as agreeing on an appointment, we are implicitly making fundamental metaphysical or ontological assumptions. These assumptions are unlikely to cross our minds unless we happened to arrange a meeting with someone – perhaps from another universe – who holds very different assumptions and who expects to meet us at a different time (perhaps even in what we consider the past). In our daily lives, we can leave such implicit assumptions unspoken, as they rarely matter. However, this is not the case in science and certainly not in interdisciplinary and transdisciplinary research.

Scientists harbor many specific assumptions, and this implicitly generates different views on shared objects and methods of research, just like our two eyes yield images that together impart extra information only when their differences are understood and integrated adequately. The most fundamental categories of assumptions relevant for science are ontological, epistemological, methodological, and normative assumptions. In addition, scientists also tacitly subscribe to assumptions that we all uphold when it comes to our understanding of culture, of what it is to be human, of moral values, and so on. We will limit our brief discussion here to the ontological, epistemological, methodological, and normative assumptions, as these have a more direct impact on research and tend to escape our notice.

Ontological assumptions pertain to what exists: forms of existence, the nature of space and time, causality, and similar fundamental issues. It might seem difficult to imagine scientists disagreeing about these issues, as that would make it difficult to understand each other or work together. Nonetheless, there are numerous instances of these differences in assumptions underlying scientific disputes. Quantum physicists, for example, do not agree among themselves about the explanations for puzzling phenomena such as entanglement, the non-local immediate interactions between two separate particles that appear to contradict long-standing ontological assumptions about the nature of causality and of space-time. An ontological assumption that biologists have had to modify since the discovery of epigenetics is the long-held assumption that DNA is the only carrier of inherited information across generations. After World War II, researchers found that environmental stress has an impact on DNA-related processes in mothers and their offspring across several generations. Since we now know that DNA transcription is affected by context-sensitive epigenetic changes, the picture of inheritable traits has changed drastically, as have our assumptions about the nature of genomes.

Related to ontological assumptions are those concerning the nature of our knowledge: *epistemological assumptions*. What can we know what we know, and what can we never know? Is it possible to obtain objective knowledge? What is the validity of our scientific laws? What is the reality of our models? Our answers to such questions implicitly shape our scientific work and understanding and can lead to heated debates between scientists. Some assume that explanations ultimately rely on the physical processes underlying the phenomena. As a result, they claim that we should all be reductionist and strive for the explanation of biological or mechanical phenomena in terms of molecular and atomic processes. Most scientists nowadays do not share this ambition of 'reducing' theories about social or natural phenomena to theories about elementary particles, as presented in the classical article on 'Unity of science as a working hypothesis' (Oppenheim & Putnam, 1958). For although it can be agreed upon that most phenomena somehow depend on such basic physical processes, at the same time, completely new properties tend to emerge from the complex and dynamic interactions between such particles. Similarly, the interactions between cells, organs, individuals, or communities tend to produce new phenomena like reproduction, consciousness, and competition. In order to understand and explain such emerging properties adequately, we need biological or social scientific theories and methods that specifically examine the levels at which these new properties arise. Consequently, physics may not have much to offer in this regard.

Not all epistemological assumptions are so fundamental as this, and yet they can still have a decisive impact on a field. Consider the fact that in many fields, scientists implicitly share epistemological assumptions about the use of specific models or research populations or text collections. Debates about such implicit epistemological assumptions have recently also had a significant impact on the interdisciplinary field of cognitive neuroscience, for example. For decades, research into the emotions of fear and anxiety has partly relied on the use of rodent models, assuming that rats' responses to threats and stress are generalizable to humans. This research has even contributed to the development of treatments and medications for phobias and disorders. But recently, an important contributor to the field admitted that he had failed to distinguish sufficiently between 'threat-induced fears', like those involving tigers and spiders, and language-dependent existential angst that humans consciously experience. Since rodents are prone to fears but not angst, this scientist admitted that assumptions about the validity of the animal model had to be revised

(LeDoux, 2014). Explicit discussions about such epistemological assumptions inevitably lead to a re-evaluation of research and the treatments based on it. Such a revision might have an impact far beyond the research lab into the clinic and even to public perceptions of fear and anxiety. Implicit as they often are, assumptions are important to articulate and critically reflect upon, certainly in interdisciplinary research.

Methodological assumptions are often related to epistemological and ontological assumptions because research methods are determined in line with ideas and expectations about knowledge and reality. In the social sciences and human sciences, for example, it is not uncommon to use a single case study, which is then analyzed in a rich and detailed narrative that strives to grasp the complexity of a particular phenomenon. Some argue that such case studies are so bound up with the particularities of their selected cases that they cannot possibly yield any insights that are more generally applicable. Others maintain that only such 'thick' descriptions can reveal the ambiguities and rich complexity that characterize real-world problems, and that it is only through such studies that we are able to understand these problems (Flyvbjerg, 2006).

In a related field, cognitive scientists have assumed for a long time that their heavy reliance on subjects from Western educated, industrialized, rich, and democratic (WEIRD) countries as their test subjects does not have an impact on the general validity of their studies of visual perception, social cognition, emotion processing, and so on. Particularly regarding vision, this methodological assumption was grounded in the theoretical assumption that visual perception has such a long evolutionary history that it is hardly 'open' to environmental influences, like those of culture. These methodological and theoretical assumptions have now been challenged, as it has been found that socio-cultural differences do have an impact on many cognitive functions, including vision (Henrich, 2020). The rejection of this methodological assumption is likely to have far-reaching consequences for how studies in the fields of cognitive science and psychology are conducted. Scientists may have to acknowledge that the validity or applicability of their test results pertains to only a small part of humanity, or they could be compelled to invest their resources in large multi-center experiments across different cultures.

Finally, *normative assumptions* inevitably play a role in science as well. Indeed, it must be acknowledged that every attempt to acquire knowledge – scientific or not – is in one way or another reliant on means-ends or cost-benefit calculations. Scientists are obliged to settle for a realistic and acceptable result while acknowledging that more comprehensive or valid results would require a much greater investment of effort (Laudan, 1986). Here again, norms are often in some sense epistemological or methodological in kind, as demonstrated by a recent position paper on the redefinition of the threshold for statistical significance – the p-value – in the context of the 'replicability crisis' in the social and human sciences (Benjamin, 2018).

More generally, normative assumptions underlie ideas about what science is and what it can accomplish, whether it is acceptable to derive ethical claims from facts of nature, how to define societal welfare in financial terms only, or whether we should develop medical knowledge (for example of contagious diseases) that might also be used in warfare – to give a few examples. Indeed, normative assumptions pervade all scientific efforts: from setting research priorities and choosing methods to interpreting results and ideas about their implementation. These assumptions are problematic when they remain implicit because this both affects critical scrutiny of such research and impedes the potential connections of such research to other interdisciplinary and transdisciplinary lines of research.

We hope that this chapter has succeeded in explaining how important it is for interdisciplinary researchers to have some grasp of the philosophy of science in order to reflect on and discuss their efforts together. These insights culminate here in the recognition of the pluralisms that pervade interdisciplinary research and the assumptions that implicitly underlie it. As interdisciplinary research involves the integration of insights from different sciences despite all their differences, understanding and recognizing these issues is a crucial starting point. When the elements discussed above remain implicit, they will often prevent scientists from collaborating with or learning from each other, since they tend to strongly influence the views that scientists hold about their work and that of their colleagues. This is something that will become apparent in the next chapter when we discuss the scientific disciplines and the historical development of interdisciplinary science.

Table 1 This table addresses four different categories of assumptions that scientists generally hold: ontological, epistemological, methodological, and normative. Interdisciplinary collaboration and integration can be facilitated by articulating these assumptions, as such assumptions guide scientific practices and reasoning, often implicitly.

Disciplinary assumptions	
Ontological	What is real and what is not, according to a discipline? What is the status of specific objects or phenomena? Are the properties of all objects, including societies, completely reducible to those of atomic particles? Is space-time absolute or curved? What is causality? Is consciousness a specific phenomenon, or is it a term applied to different phenomena? Is marriage a contract or only a speech act?
Epistemological	What is knowledge and what can be known? Is our knowledge a reliable 'mirror of reality', or is it something that humans have constructed, or is it as a mere instrument that works, or not? What is the validity of scientific laws and how reliable are our models? Is quantum physics the most fundamental science, to which all other sciences are reducible in some way or another? Are we applying concepts for human cognition to our animal models?
Methodological	How can we adequately develop robust knowledge about a phenomenon, what research strategy should we apply? What experiments or tests should we apply and what objects or subjects should we use in these? Are only large numbers of quantitative data reliable or do in-depth interviews also yield relevant knowledge? Is our group of test subjects sufficiently representative? How do we handle the methodological pluralism involved in interdisciplinary research?
Normative	What problems or questions should we dedicate our resources to? When may we be satisfied with our research, how many tests should we do or data should we gather? What role should the potential applicability of results play in our decision to conduct a study?

3 From disciplines to interdisciplinarity

3.1 Knowledge formation and the emergence of disciplines

In the previous chapter, we examined the process of producing scholarly insights and knowledge by explaining the steps of the cycle of scientific research, which carried us from theories and laws to the deduction of testable predictions. Testing those predictions produces results, which must be processed and evaluated in order to yield relevant and adequate observations. In many cases, these observations invite us to reconsider and adjust the theory and/or law that we started with – inviting us to move through this science cycle once more with the now improved set of theories and laws. The chapter ended with a discussion of the methodological and theoretical pluralisms that are typically involved in a proper investigation of a complex phenomenon – such as the relationship between alcohol and aggression – given that so many interacting factors potentially influence it. In addition, we looked at the many assumptions that implicitly determine the way we conceive the phenomena we are investigating, the kind of knowledge we aim to produce, the methods we believe to be best suited for this task, and the norms that govern our expectations. This foundational analysis of the scientific research process will support us when conducting interdisciplinary research, as it also enables us to understand how disciplines operate, what the differences between disciplines are, and how we can integrate disciplinary contributions to develop optimal solutions to a scientific problem like alcohol-induced aggression.

Thus far, the notion of 'disciplines' has been largely absent from our analysis, even if we started our introductory chapter with Popper's observation that 'problems may cut right across the boundaries of any subject matter or discipline' (Popper, 2002). It might seem somewhat surprising that a potential mismatch could exist between a problem and a discipline, as it suggests a disparity between the domain of a problem and that of a discipline. For isn't the defining feature of a discipline its focus on a particular domain, as when biology solves questions about flora and fauna and theology addresses religion and religious phenomena? And what about the specific instruments and artifacts that we often associate with a discipline – such as the biologist's microscope, the astronomer's telescope, the historian's archives, and the economist's computer model? In this chapter, we will discuss the various characteristics that have historically – and hence somewhat accidentally – become associated with scientific disciplines. This perspective on disciplines builds on the philosophical perspective described in the previous chapter aimed at showing how the interdisciplinary integration of disciplinary perspectives is not only possible but, in many cases, the appropriate approach.

3.2 Pre-disciplinary approaches to the organization of knowledge

The word 'discipline' derives from the Latin verb 'discere' meaning 'to learn', with 'disciplina' referring to 'instruction'. Scholars and philosophers have over time defined scholarly disciplines in quite different ways. Popper's contention that the scientific approach to problems should defy the notion of distinct 'subject matters' located conveniently within neat 'boundaries' is a critique of the influential traditional view of scientific disciplines, which holds that they should be organized around distinct domains of reality. Since this traditional view has left its mark on the organization and structure of science to this very day, we pause briefly here to clarify that view and to explore the emergence of scientific disciplines.

The traditional view of scientific disciplines is based on the ontological assumption that reality is made up of kinds of things that are fundamentally different from each other. At first sight, this assumption seems to make sense, as it corresponds to our everyday perceptions of the world around us. The characteristic properties of material objects such as stones and stars appear very different from those of biological objects such as plants and animals: the latter grow and move, respond to environmental stimuli, and die after some time, while stones and stars do not (as far as we can perceive in our short-term observations). Similarly, the phenomena that psychologists investigate – emotions and behavior – are fundamentally different from the groups and processes that social scientists investigate, which in turn cannot be compared with the series of numbers or solids that mathematicians study. Recognizing these differences, the traditional view contends that each of those phenomena or objects requires a correspondingly distinct set of methods, which would then yield equally distinct results and theories. Today, this differentiation of phenomena, methods, and theories comes across as quite strange, especially for those involved in interdisciplinary research like social psychologists. Instead of ridiculing this traditional view, though, it is useful to consider the motivations behind it.

We noted in the previous chapter that science generally builds upon two processes: one that involves gathering empirical data or information via some form of perception, and another that entails circumscribed forms of reasoning to adequately draw conclusions from perceived sets of data. The very productive and influential scientist and philosopher Aristotle (384-322 BC) developed an analysis of how scientists systematically organize and categorize their observations of natural phenomena in order to reason about their causal structures (Keestra, 2000). Aiming to distinguish scientific knowledge from other ways of thinking, in particular superstition, he emphasized the role of logical reasoning. However, Aristotle contended that drawing robust conclusions about a phenomenon or object after sound reasoning of our observations of it depends to a large extent upon our ability to offer a proper definition of it in the first place. Consider, for example, a situation in which the definition of 'aggression' is so vague that scientists can individually decide whether they only include violent behavior or also count stereotypical, sexist, and racist jokes as violence. Such a dissensus will make the exchange and integration of insights not only challenging but also close to impossible and have a negative impact on their experiments and methods. In addition, it will impede the implementation of insights in behavioral therapies or professional codes of conduct. Similarly, unclear or varying definitions of free will, consciousness, or religion have negative consequences in ethics, psychology, and anthropology.

A conceptual lack of clarity and its consequences are largely absent from the field of mathematics, which has been admired for its precision from antiquity to the present day. Mathematicians can more or less create their own 'subject matter' by providing clear and distinct definitions, axioms, and postulates. Starting from precise definitions of a point and a line, supported by a small number of postulates, they have developed, using logical reasoning, an extensive geometry consisting of circles, pyramids, spheres, and other objects. The progress of this field of knowledge and the growth of insights it has produced convinced many scholars since antiquity that the structure of this discipline had a superior value. Moreover, its methodology not only made mathematics appealing as an intellectual endeavor, but it was also instrumental in providing solutions to real-world problems such as calculating the circumference of the earth and digging a kilometer-long tunnel for water transport. Unsurprisingly, then, Aristotle and many philosophers and scholars from different disciplines saw mathematics as an exemplar because of its 'axiomatic structure'. Naturally, they tried to apply its structure to other domains of reality, arguing that scientific knowledge could be obtained even outside of mathematics by using adequate definitions and the right procedures of reasoning. This effort continues to this day, as demonstrated by more recent attempts to axiomatize fields as varied as quantum physics, game theory, economics, artificial intelligence, and evolutionary biology. Discussing a recent axiomatic theory of evolution, for example, Bocci and Freguglia point out that a logical formulation of its contents makes it easier 'to explore all their implications and to look for all possible verifications' (Bocci & Freguglia, 2006). Hence, precise and consistent definitions of concepts and hypotheses facilitate both the reasoning and the empirical components of the science cycle.

These developments show that axiomatics can be attractive as a method even if one does not subscribe to the ontological assumption that there are different domains of reality. In that case, axiomatics is applied as an instrument to organize knowledge systematically and consistently, irrespective of how it was produced. Weingart further argues that, preceding the period in which disciplines came to play an increasingly important role in the growth and organization of knowledge, scientists in the 17th and 18th centuries were particularly engaged in systematizing and classifying existing knowledge. This development led at the time to numerous encyclopedias and taxonomies of knowledge domains akin to the classifications of plants and animals (Weingart, 2010). These compilers' works relied on epistemological assumptions pertaining to the nature of knowledge, its different domains, and its suitability for such types of organization. In addition to this engagement with the organization of knowledge, from the 15th century onwards there was a growing interest in the systematic production of new knowledge, which gave rise to the emergence of disciplines as we know them today.

3.3 From knowledge formation to disciplines

In the previous chapter, we emphasized that science consists of two processes that must be combined: perception and reasoning. Induction and deduction became the two main forms of reasoning involved in the science cycle, as they connect multiple scientific statements in a logically consistent and meaningful way. Perception, the other process, can occur in multiple forms as well, ranging from direct sense perception – such as observing the number of plants in a designated area or the forms of interaction between people – to more specialized and mediated forms of perception that rely on the use of technologies, such as fMRI-scanning of human bodies or the Large Hadron Collider in particle physics.

Although perception and observation have been part and parcel of science since antiquity, several centuries passed before scientists in Europe were expected to produce and control them in more specific ways. Indeed, the notion of 'experimentation' as methodologically produced observations emerged only in early modern times. Although experiments were already designed by scientists such as Galileo (1564-1642), who tested his notions of gravity, a more precisely defined experimental method was championed only in Francis Bacon's 1620 treatise *Novum Organum* (New Method). For Bacon, a systematic survey of existing knowledge could be used to discover gaps in human knowledge, but he argued that the next step should be to determine what new knowledge is needed to fill them. Indeed, he believed that the growth of knowledge rather than simply the organization of what we already know should be the aim of scientific work. He offered the idea that experiments allow for the controlled and repeated observations of the same phenomenon – described, if possible, in mathematical terms, echoing Galileo's influential adage that 'the Book of Nature is written in the language of mathematics'. Using induction, experiments producing a series of observations would lead to new knowledge that revealed natural laws, proportions of substances, or other insights into the natural world that escape our usual perception.

Influenced by Bacon and others, scientists in the 17th and 18th centuries began favoring empirical experimentation over more traditional encyclopedic and classificatory endeavors. Emphasizing the growth of knowledge through the use of experiments and supported by instruments such as microscopes and thermometers, science became an activity that no longer required erudition and knowledge acquired from books. Indeed, the so-called Scientific Revolution – dated roughly from the Renaissance to Newton's publication of his seminal book *Principia* in 1687 – can be seen as a form of democratization of science. Since observation and experimentation often involved the use of instruments, it also required collaboration with artisans outside of educational institutes. Associated with these developments, scientific academies and societies were established in several countries, located outside the universities and turning science increasingly into a collective endeavor relying on the mastery of multiple techniques and practices (Cohen, 1994).

This thumbnail sketch of the rise of modern science suffices to account for the emergence of scientific disciplines as we think of them today, given that such disciplines structure the organization of our universities and are tasked with training pupils or 'discipuli'. It confirms the historical and contingent nature of that organization, which remains to some extent in flux. Traditionally, disciplines functioned largely as units for archiving existing knowledge, and the purpose of the university was to transmit that knowledge. Although the establishment of universities in Europe began in the 12th century – while the Indian university of Nalanda was already founded in 427 – these medieval universities were focused on theology, law, and medicine. They certainly did not engage in the kind of empirical experimentation that has become standard in modern scientific disciplines, which entails critical analysis of existing knowledge combined with the use of instruments and techniques for controlled observations and experimentation, technical mathematical language to describe the results, and inductive reasoning to produce new insights. Because the different sciences – biology, psychology, mathematics, etc. – were assumed to cover separate domains of reality and associated with distinct domains of knowledge needing systematization, they emerged in modern form in the 19th century as separate 'disciplines' in universities. The traditional training imparted by universities, which had largely consisted of the faithful reproduction of existing knowledge in a certain domain, began to transform into a form that emphasized preparing and assessing students in their ability to contribute independently to the growth of knowledge. This is the current organizational structure of the university that you – as a student, teacher, and/or researcher – are a part of. This short history hopefully provides you with sufficient context to grasp the challenges this structure poses to the types of interdisciplinary research we describe in this book.

3.4 Discipline: a plural concept

The foregoing brief account of the emergence of scientific disciplines in the modern university illustrates the multiple ways in which disciplines have been defined. Some define disciplines based on the notion of separate domains of reality, while others present disciplines as representing well-defined fields of knowledge. More recently, disciplines have come to be considered 'production and communication systems' that are responsible for producing and communicating knowledge (Stichweh, 2001). This last connotation of disciplinarity resonates with the philosopher and historian of science Thomas Kuhn's emphasis on the growth and organization of knowledge. Treating the history of science as a mixture of growth and stagnation of knowledge, he coined the term 'paradigm' to characterize a particular scientific discipline at a given time (Kuhn, 1970 [1962]). A paradigm is a broad set of concepts, assumptions, values, theories, methods, instruments, and even social practices shared by a particular community of scientists. This implies that social factors, values, and habits shape a discipline's boundaries and contents, such as a community of colleagues who share practices and ideas about a particular domain. Thus, according to Kuhn, we cannot claim that a scientific discipline is solely determined by the definitions, axioms, postulates, etc. that determine its contents or by the methods used to investigate these definitions, axioms, etc.

This broader notion of a scientific discipline means that it is possible to have different sub-communities of scientists operating simultaneously within one and the same discipline, each subscribing to different paradigms: for example, Marxists and structuralists within the social sciences, those who expect string theory to solve the unification problem in physics and those who do not, traditional economists versus 'doughnut economists' who argue that economies should neither extend beyond their ecological ceilings nor fail to meet certain basic social aims. Such differences imply that the term 'discipline' has inevitably lost some of its appeal given the lack of consensus about its contents, methods, practices, and other distinguishing features. According to some, disciplines even have lost their 'disciplinary' nature. Such rifts occur in all disciplines, as the history of science demonstrates. Famously, early modern astronomy was for some time divided between one group of astronomers who grounded their work in a Ptolemaic, geocentric cosmology and another group adhering to the alternative Copernican and heliocentric paradigm. Gradually, most astronomers became convinced of the theoretical and empirical strengths of the latter and as a result abandoned the Ptolemaic paradigm.

Scientists such as Galileo, Einstein, Darwin, Marx, and Freud brought about similar 'Copernican' revolutions in physics, biology, economics, and psychology by presenting their alternatives to the paradigms existing at the time. Their revolutionary ideas were initially accepted by only a limited group of colleagues, whereas many others remained loyal to the old 'paradigms' and tried to defend their disciplines against what they considered 'attacks'. After a while, however, the shortcomings of the traditional paradigms and the validity of the new ones became more widely recognized. In some cases, a so-called paradigm shift occurred within those disciplines, followed by a period of 'normal science' in which the new paradigm became established. In other cases, the two competing paradigms merged, or two different paradigms continued to co-exist.

This history makes it clear that it is impossible to call something a discipline simply by combining a clear set of criteria with a specific field of inquiry and education. Therefore, we define a discipline here as a field of systematic research of a loosely definable object domain and a corresponding body of accumulated specialist knowledge and expertise expressed in a set of theories, concepts, methods, and relevant assumptions. Each discipline has discipline-specific terminologies, technical language, and expertise regarding research instruments and artifacts such as telescopes, experimental instruments, and linguistic corpora. Furthermore, a discipline has several social and institutional manifestations such as dedicated conferences, journals, professional associations, and university educational programs. The last criterion is especially important because the handing down of a discipline from generation to generation calls for a new generation to be trained and prepared in that discipline. Summing up, a discipline can be said to represent a particular *perspective* or *lens* that both facilitates and constrains how we perceive and understand reality (Repko & Szostak, 2017).

This somewhat loose definition of a discipline is consistent with the notion of theoretical and methodological pluralism that we discussed in Chapter 2. It also allows for the co-existence of competing paradigms within a discipline. However, this makes it difficult for us to draw clear-cut boundaries between disciplines. Once we recognize these complexities, it becomes easier to understand how to facilitate interdisciplinary and transdisciplinary collaborations. Some collaborations may be based on shared concepts, others on shared methods or theories, and still others might share values about what constitutes useful knowledge. As we shall see in Chapter 5, some collaborations may succeed in achieving interdisciplinary integration even in more than one way. Importantly, scientists should be aware that their collaborative efforts may leave differences intact regarding their views on the components of reality (ontology), on true and reliable knowledge (epistemology), and on what are important and what are less significant questions (including moral concerns). The challenge for interdisciplinarians is to be aware of such differences and to articulate and communicate them in order to prevent them from impeding the research. This will be addressed explicitly in Chapter 6 on the interdisciplinary research process.

3.5 Increasing specialization and isolation of disciplines

Academic disciplines reflect how scientists produce and organize their knowledge, which can change over time given the shifting focus of interest within disciplines and the flexible boundaries between them. Of course, some disciplines may exhibit more internal consensus than others regarding the research domain, the accepted body of knowledge, and the methods to be used. For example, mathematicians and astronomers tend to be more in agreement within their discipline than is the case with sociologists, cultural analysts, and philosophers. Similarly, textbooks and academic programs in astronomy show less divergence than their counterparts in sociology. Given the fluidity of disciplines, multiple classifications of the sciences are possible, none of which can satisfy all purposes (cf. Szostak, 2004). If we look at the Dutch education system, it distinguishes between the so-called alpha, beta, and gamma domains, which roughly correspond to the humanities, the natural sciences, and the social and behavioral sciences. But even with this division, there is no agreement on where to place philosophy – which in some sense is a meta-science relevant for all disciplines – or whether anthropology belongs to the humanities or the social sciences.

Furthermore, there is a lack of clear boundaries between disciplines as a result of their increasing specialization, which has fortunately coincided with an increase in cross-fertilization. As described above, becoming a scientist in more recent times implies becoming part of a 'communication and production system' that requires extensive education as well as training to prepare scientists for special conferences, journals, and other means of collegial communication. This process leads to the separation of members of a discipline from their counterparts in other specialties, as the sociology of group formation has proven. In the case of science, there is a further element at work. For research purposes, it is often necessary to isolate a particular object or feature from numerous other factors to which it might be related. For example, experimenters investigating the relationship between alcohol and aggression might develop an experimental paradigm in which the time of the year does not matter for test results. Interpreting a particular religious symbol requires familiarity with its traditional contexts but not necessarily the physical geography a particular religious sect inhabits – unless further research proves this assumption wrong. Scientists employ this method of isolation in order to more easily reach results that can lead to scientific laws and precise interpretations, rather than considering a myriad of determining factors. This fosters increasing specialization within disciplines, as can be seen by the numerous specialized sub-disciplines or branches within each scientific discipline. Figure 6 below shows the multitude of subjects within the field of biology.

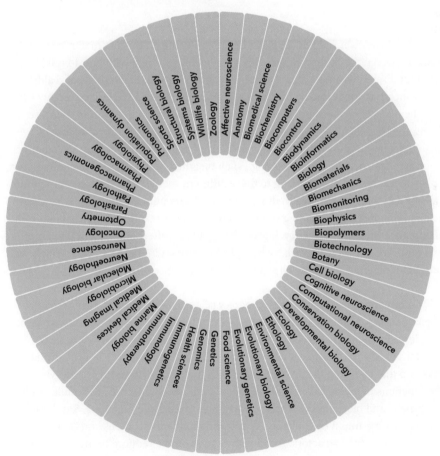

Figure 6   The many branches of biology

Some of these branches are classified according to the level at which the subject is studied, such as molecular biology, cell biology, or ecology. Other sub-disciplines are characterized by the type of organism studied (e.g., botany and zoology) or their focus on a practical application (e.g., conservation biology). The same trend can be seen in other disciplines as well.

The ingredients included in our definition of a discipline will apply to most if not all branches. From affective neuroscience to zoology, each has a loosely definable object domain and a corresponding body of accumulated specialist knowledge and expertise involving theories, methods, and relevant assumptions. Depending on how mature the discipline is, the number and size of the specific conferences, journals, and PhD programs focusing on its domain may increase. Correspondingly, students trained in a particular sub-discipline will have gathered highly specialized theoretical and methodological expertise, making them typically less employable in other sub-disciplines even within the same field. Nonetheless, disciplinary specialization and isolation occur in parallel with a development in the other direction: collaboration and interdisciplinarity.

3.6 Unification of science or collaborations between disciplines

Ever since scientists and philosophers began reflecting on the distinction between branches of science or disciplines, they have been concerned with how to unify the different disciplines into a single theory encompassing all the domains of reality. This concern is motivated by ontological assumptions about the unity of reality that risks being lost if scientific disciplines investigate parts of it in isolation. It is also motivated by epistemological assumptions about the necessity for coherence of distinct bodies of knowledge, and by normative assumptions about the socio-political implications of scientific insights (e.g., such insights should not lead to contradictory policy measures). Corresponding to these different motivations, the attempt to unify science has been approached in different ways. As mentioned above, early modern attempts to offer a systematic inventory of the sciences resulted in efforts to present all scientific knowledge in encyclopedic form. A very different and radical attempt was proposed a few centuries later by logical positivists in the early 20th century who sought to integrate and unify all sciences via the practice of reductionism.

Still an important doctrine in contemporary science, reductionism posits that a phenomenon – or a theory that explains it – can be redescribed in terms of its most simple and basic elements. It contends that we can understand the properties of a material body, for example, once we redescribe it in terms of the molecules that make up that body and the properties of those molecules. According to this view, the functions of biological organisms can be understood once we are familiar with the genetic and biochemical processes that determine these functions. Similarly, a social process might be explained by referring to the motivations and actions of contributing individual actors. As all phenomena can be reduced to their smaller components, reductionists presume that knowledge of the basic elements of reality would enable the reduction of all knowledge to a foundational form of scientific knowledge. Thus, all scientific knowledge can eventually be reduced to particle physics, the branch of physics covering the most elementary components of matter.

What reductionism cannot explain, however, is the obvious fact that a complex and dynamic system such as a society or an organism – or even certain molecular compounds – displays properties and behaviors that cannot be observed at the level of their components. This is because the components of such a system interact with each another, bringing about entirely new properties that are called emergent properties. In addition, such systems typically interact with their environments, which also generates new features and developments. For example, the properties of water, including its fluidity or phase transitions caused by temperature change, cannot be derived from the properties of either oxygen or hydrogen; they emerge only as a result of their interaction. An organism interacts with and adjusts to its environment, engendering differences over time between individuals of the same species. Similarly, group behavior, including the use of language, cannot be reduced to the aggregation of the behaviors of isolated individuals: democracy or warfare, for example, give rise to behavior and cognition that individuals would not exhibit on their own. For these reasons, a reductionist explanation is not viable for most phenomena pertaining to the world of living creatures.

Around the same time that reductionists were attempting to unify the increasingly specialized and isolated disciplines, a very different approach was proposed, one that can be considered the origin of interdisciplinarity. In 1925, at a time when general education programs were also established, the Social Science Research Council (SSRC) in New York started a funding program to facilitate collaborations between the multiple social scientific disciplines it oversaw with the aim of encouraging problem-focused research (Frank, 1988). The impetus behind this interdisciplinary program was the general dissatisfaction with the absence of commonality between different disciplines such as anthropology, political science, and sociology in terms of concepts, theories, and methods as well as the lack of interest shown in each other's results. To combat this problem, the SSRC board supported joint, interdisciplinary projects that it hoped would lead to an understanding of the dynamic interactions between cultural symbols, their use in political struggles, and the impact of this on the formation of groups.

Although much less ambitious than the reductionist unification program, this type of targeted interdisciplinary collaboration between disciplines has over time become increasingly common. In the next chapter, we focus on the nature and dynamics of interdisciplinarity and consider some of the drivers behind the move towards interdisciplinarity.

4 Interdisciplinarity and its ongoing developments

4.1 Weaving disciplines: multidisciplinarity, interdisciplinarity, and transdisciplinarity

Disciplines are strange beasts, as the previous chapter revealed. We settled on a pragmatic definition of the word 'discipline' as a field of systematic research on a loosely definable object domain and a corresponding body of accumulated specialist knowledge and expertise. We further specified that a discipline encompasses a set of theories, concepts, and methods (sometimes even somewhat inconsistent or competing) as well as the assumptions and norms that govern those. In addition, disciplines have their own specialized language, instruments, and tools as well as social and institutional manifestations, including educational programs and journals. We noted that this unwieldy species known as disciplines tends to become increasingly specialized, isolating individuals by virtue of not only their languages, tools, and research objects but also their curricula, training programs, and conferences.

Despite the segmentation generated by these characteristics, efforts ensued to connect the different disciplines, aligned with the notion of an 'interdisciplinary' approach that was introduced in the 1920s. This chapter is dedicated to the major factors driving interdisciplinarity and the ongoing developments in this area. Before delving into this, however, we should first address the question of how we can connect disciplines. Combining disciplines when solving a particular problem can take quite a few different forms. This pluralism partly explains the disparate terminology used to label distinct forms and activities. In their list of 'modes of disciplinary combination', O'Rourke et al. include cross-disciplinarity (in which one discipline is privileged over other subservient specialties), intra-disciplinarity (in which different strands of a single discipline are integrated), multi-disciplinarity (elaborated below), pluri-disciplinarity (connection of more or less related disciplines), post-disciplinarity (disciplines' loss of categorical status), supra-disciplinarity (an umbrella term for all forms of scientific collaboration, such as poly-disciplinarity), and trans-disciplinarity, which we explain below (O'Rourke et al., 2019). Klein offers an even longer list of modes according to which disciplines can be combined, integrated, juxtaposed, coordinated, sequenced, etc. (Klein, 2017).

For the purposes of this book, we limit ourselves to the three most common modes of disciplinary combination, as these allow us to reflect on the foundational characteristics that students, teachers, and researchers might consider when solving complex problems, regardless of whether they are practical or more theoretical problems. Figure 7 depicts the different relationships between the disciplines based on the three different modes of multidisciplinarity, interdisciplinarity, and transdisciplinarity:

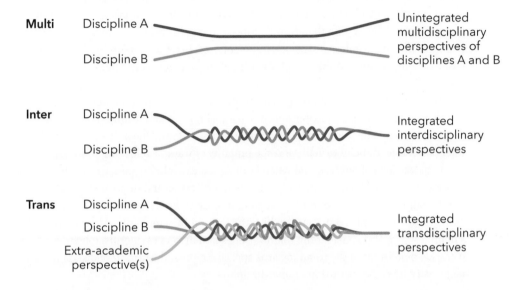

Figure 7 Multidisciplinarity, interdisciplinarity, and transdisciplinarity illustrated, highlighting the absence or presence of the integration of perspectives

Two features stand out in this figure. First, the top and middle images include only two lines (representing Discipline A and Discipline B), while the bottom image has a third line that represents extra-academic insights. This concurs with an increasingly accepted understanding of transdisciplinarity as developed in Europe (Hirsch Hadorn et al., 2008). Transgressing or transcending disciplinary boundaries altogether, transdisciplinary research incorporates relevant stakeholders in a project, since they bring relevant experiential knowledge to the table regardless of whether they have academic training. When a project aims to produce insights or social interventions that are meant to be 'socially robust' and viable beyond the research lab or clinic, the inclusion of extra-academic stakeholders is usually key to ensuring that the research and the solutions stemming from it are relevant and acceptable to affected citizens, patients, clients, or other stakeholders.

Second, the absence or presence of integration between lines is noteworthy. The crucial difference between multidisciplinarity and interdisciplinarity is the lack of integration between disciplinary contributions in the former. It is one thing to investigate climate change, for example, or the relation between alcohol and aggression from different disciplinary perspectives. Yet it is quite another to offer an explanation, prediction, or intervention that integrates the relevant components that disciplines can contribute to useful interventions in combating climate change or alcohol-induced aggression. In both cases, research might suggest multiple kinds of interventions, each of which also has an economic impact in terms of cost/benefit ratios. Choosing the most effective intervention depends on which factors we include in our considerations. This choice must also hinge on the social and political impact of the intervention in order to prevent relevant stakeholders from resisting or not complying with the intervention. Integrating insights into a final choice of the most effective intervention therefore requires a more pluralistic notion of the 'costs' related to the effectiveness of any one intervention. This is a matter of optimal choice based on a recognition of the complex nature of impact and acceptability.

Clearly, integrating different disciplinary contributions is more challenging than simply presenting these contributions in a parallel relationship (multidisciplinarity). Yet socially robust decisions call for an integrated approach. One of the reasons that integration is so difficult to achieve is the plurality of ways of achieving it, which involves decisions about how and when to integrate disciplinary perspectives for a given purpose. Chapter 5 will explain in greater detail several important forms of integration that might be implemented in interdisciplinary research. At this point, however, we turn our focus to the nature of interdisciplinarity. After offering a definition, we also consider the main varieties and drivers of interdisciplinarity, while keeping in mind the trend towards specialization and isolation of disciplines identified earlier as a backdrop to interdisciplinarity.

4.2 Interdisciplinarity: a definition and an explanation

After the term was introduced in the 1920s, interdisciplinarity grew steadily in the research landscape as well as in higher education. Nonetheless, it took a while to establish the precise meaning of interdisciplinarity, even though the concept was already being realized in many forms. In the 1930s and 1940s, area studies were established, focusing for example on the Pacific region or Latin America. Several intelligence, defense, and policy projects during World War II and the post-war reconstruction period required intense collaborations across not only disciplinary boundaries but also across sectors of government, the military, and industry. This development supported the growing interest in this new kind of research, posing a challenge to those universities that could not easily accommodate boundary-crossing collaborations. New fields such as women's studies, environmental studies, and urban studies emerged in the 1960s and 1970s. Responding to the continuing impediments to interdisciplinary research and education, a 1970 seminar co-sponsored by the international Organisation for Economic Cooperation and Development (OECD) focused on interdisciplinarity and the 'problems of teaching and research at universities'. Contributors noted the difficulties they encountered when bringing together insights across disciplines in their academic environments. The book that emanated from the seminar argued that special arrangements should be made to facilitate exchanges that are both interdisciplinary and international (Apostel et al., 1972).

Although universities, journals, conferences, etc. are still predominantly organized according to the logic of separate disciplines, interdisciplinary collaboration between disciplinary experts has become increasingly common and widely endorsed. Since such collaborations are generally specific to particular problems, research questions, or case studies, definitions of interdisciplinarity must allow for the many varieties in collaboration they require. Aiming not to exclude particular manifestations, Klein and Newell define interdisciplinary research as 'a process of answering a question, solving a problem, or addressing a topic that is too broad or complex to be dealt with adequately by a single discipline or profession [...] [that] draws on disciplinary perspectives and integrates their insights through construction of a more comprehensive perspective' (Klein & Newell, 1997).

An even more influential definition emerged from the US-based National Academies of Sciences, Engineering and Medicine (NASEM), published in its state-of-the-art report titled *Facilitating Interdisciplinary Research*:

> Interdisciplinary research is a mode of research in which an individual scientist or a team of scientists integrates information, data, techniques, tools, perspectives, concepts, and/or theories from two or more disciplines or bodies of specialized knowledge, with the objective to advance fundamental understanding or to solve problems whose solutions are beyond the scope of a single discipline or area of research practice. (National Academies of Sciences, 2005)

Both definitions point out that the results of interdisciplinary research are more comprehensive than monodisciplinary results and produce solutions that are outside of the purview of a single discipline. To produce such comprehensive interdisciplinary results, the integration of one or more key ingredients from different disciplines is essential.

What these two definitions do not shed light on, however, is how these ingredients are related to the components of the cycle of science analyzed in our previous chapter on philosophy of science. This is what we explain in the next chapter, where we describe the different varieties of interdisciplinary integration in terms of the steps and components of the cycle and the ways in which several of them might be applied in parallel for a single project. First, however, we need to consider the dimensions of variation in interdisciplinary research and education.

4.3 Variations in interdisciplinary research

The previous section distinguished different ways of combining perspectives, whether they be academic, extra-academic, or a combination of both. Thus, interdisciplinary research can take on different shapes as a result of the choices made in a particular case or context – choices such as whether integration should take place or whether extra-academic perspectives should be included. We turn now to the dimensions that are affected by such decisions. Together, these dimensions demonstrate the pluralism of interdisciplinarity.

Dimension 1: Narrow versus broad interdisciplinarity

Disciplines are increasingly characterized by theoretical and methodological pluralism, as we showed in the last chapter. This development has led to the emergence of sub-disciplines that differ from each other, such as affective neuroscience and zoology, or ecology and molecular biology. However, these sub-disciplines are still relatively close to each other compared to the combination of biology and theology, or anthropology and quantum physics. When disciplines are relatively close to each other epistemologically, their relationship is one of what we call 'narrow interdisciplinarity'. Conversely, if disciplines differ significantly in terms of research objects, theories, and methods, their integration entails 'broad interdisciplinarity' (Newell, 2007). Generally speaking, a broad interdisciplinary research project is more challenging because not only the research objects but also the methods used to investigate them might differ significantly. In addition, implicit assumptions in those disciplines can create obstacles to collaboration. For example, anthropologists are used to conducting participatory observation of a culture and therefore tend to reflect on their own positionality and cultural biases that may influence their interpretations of phenomena and related symbols, actions, and language, while physicists typically conduct laboratory experiments in which their personal influences are excluded as much as possible, or they approach problems using a formal, unambiguous language such as mathematics. Although both disciplines face problems of interpretation, interpretations in quantum physics are of a different nature than in anthropology. If these two disciplines were to collaborate on different conceptualizations of consciousness and the potential role of quantum phenomena therein, for example, they need to make explicit their relevant differences and implicit assumptions in order to achieve interdisciplinary integration.

Dimension 2: Number and relevance of disciplines involved

According to an old adage, everything is connected with everything else. One could conclude from this that eventually all disciplines should weigh in on a particular research project. However, this approach is not feasible because from a practical standpoint, most objects can only be investigated using a few disciplines. Furthermore, since not all factors have an equally significant impact on an object, associated disciplines differ in relevance when aiming to explain or understand it. Indeed, when we are aiming to explain or predict a particular property of a phenomenon, we must identify those of its components that contribute specifically to those properties while leaving out of our investigation all others, and therefore, assessing their relevance can be useful (Craver, 2007). When explaining aggressive behavior, although a person cannot be aggressive if their lungs do not function, for most healthy persons their lung properties are not relevant for an investigation into aggression. Economists differ in opinion as to whether national differences in productivity are determined by culture or by the way in which nations handle property rights (De Soto, 2003).

Given that an interdisciplinary project must be manageable, the number of disciplines involved should be limited. Two questions follow from this. Which disciplines should be included based on their relevance to the nature or the variability of the phenomenon? And what factors can we easily manipulate if we aim to intervene in the phenomenon? When investigating the relation between alcohol and aggression, for example, genetics might be helpful in explaining their relation. But if the aim of the research project is to devise an intervention, it may make more sense to include a psychotherapist on the team instead of a geneticist. As this example illustrates, relevance can be determined in several ways, but decisions must be made about which disciplines are needed most. Moreover, even after choosing the disciplines to be included in our interdisciplinary research, we may at some point need to reconsider this decision if additional disciplines turn out to be relevant or if some disciplines lose their relevance to the research.

Dimension 3: Case-based and theory-driven interdisciplinarity

The last two sources of variation merit a more detailed analysis. As we shall see below, interdisciplinary research is often conducted in response to complex problems such as the migration crisis, climate change, and the challenge of traveling to Mars. If such problems are framed as a case study, we can use the characteristics of the case to circumscribe a research project. Global warming, for instance, is a complex phenomenon determined by multiple, differing dimensions on both the global and local scale. Hence, it requires joint efforts by multiple disciplines and many contributing scientists, as reports from the Intergovernmental Panel on Climate Change testify. However, when narrowed down to a specific case, decision-making becomes more manageable. For instance, tackling the impact of global warming on a northern harbor city such as Rotterdam requires input from demographers, political scientists, trade economists, and oceanographers. This combination of experts, though, might differ from an interdisciplinary team developing mitigation strategies against the impact of global warming on the Himalayas, which might require instead the input of glaciologists and religious scholars. In any event, both cases also require the integration of meteorological and demographical data.

There are those who contend that interdisciplinary research should mostly be aimed at solving cases, while others see research as a means to develop general theories and laws (Krohn, 2010). Even though this distinction might hold in many instances, it is not necessarily relevant if interdisciplinary researchers don't aim to develop general scientific insights in a given case. As noted in the previous chapter, interdisciplinarity often emerges at the border between different disciplines or sub-disciplines. In these instances, the theories of one discipline are applied to another discipline's method or object field, perhaps at first in an analogical sense. Fundamental and generally applicable insights might result. Game theory, for example, was first developed in economics but has since been applied to social psychology, ecology, and even oncology to explain forms of collaboration or competition between subjects, species, and cells. Productive applications of a theory across disciplines are a valuable interdisciplinary outcome that is more generally applicable than results of an interdisciplinary case study.

Another way to consider how interdisciplinary projects might vary is to distinguish interdisciplinary research focused on a specific problem from research that aims to develop a general theory or law. Related to this is the distinction between interdisciplinarity as contributing to 'practical problem-solving' and interdisciplinarity that contributes to the growth of knowledge without any practical applicability. These two distinctions are by no means identical, even if they refer to dimensions that are to some extent comparable. They face the common challenge of circumscribing an interdisciplinary project in order to keep it within reasonable bounds. Focusing on a particular case with its spatial, temporal, and other constraints helps to limit and manage a project. It also helps to decide which disciplines are most relevant, what datasets bear on it, and what kind of stakeholders might be good to interview.

Dimension 4: Levels of integration

In her seminal monograph *Interdisciplinarity: History, Theory, and Practice*, Julie Thompson Klein distinguishes four basic kinds of interdisciplinary 'interaction' that differ in terms of the degree of integration between disciplines and collaboration (Klein, 1990).

First, one discipline can merely borrow a particular ingredient such as an analytical tool, a concept, or even a theory from another discipline without individuals necessarily engaging in collaboration. Such was the case initially when game theory was borrowed from economics to describe and explain ecological phenomena.

Second, solving a specific (practical) problem might require a more wide-ranging yet clearly defined collaboration, as was the case with the Apollo space project. Nonetheless, the disciplines involved in this technological and practical project were not necessarily interested in further integration of their underlying bodies of knowledge, methods and theories.

Third, possibly resulting from previous shared research projects, two disciplines might overlap increasingly with regard to topics of research and methods. By offering multiple contributions, they become increasingly aligned along several dimensions. Such developments gave rise to neuro-anthropology, which investigates interactions between culture and brain development and function, and digital humanities, which addresses questions in humanities using computational and other digital methods.

The fourth kind Klein distinguishes is when the blurring of boundaries between disciplines even lead to emergence of a new 'interdiscipline' characterized by its own new body of laws, set of methods, basic questions and concepts, and mechanisms such as specialized journals, professional societies, and textbooks, all characteristics of a discipline. For example, cognitive neuroscience has been considered a separate inter-discipline since 1970, when decades of research resulted in an increasing number of elements of psychological and cognitive science with neurobiology and neuroscience.

The four kinds of interdisciplinarity listed above is by no means exhaustive. More decisions must be made by an interdisciplinary project team, each of which will have ramifications for the nature of a given project, its complexity, its temporal and spatial constraints, the applicability of its results, the potential audiences, etc. (cf. Lyall, 2008). In the following section, we examine more dimensions that might affect the nature of interdisciplinarity but also contribute to its growing importance.

4.4 Drivers of interdisciplinarity

To reiterate, dissatisfaction with the increasing specialization and isolation of scientific disciplines prompted the first efforts to bridge them in the 1920s. Eighty years later, the widely recognized definition of interdisciplinary research in the aforementioned National Academies report identified what drives the call for interdisciplinarity today. In this section, we briefly present four such drivers before turning to the issue of complexity and wicked problems, which also require an interdisciplinary approach. When looking for a new interdisciplinary research object, it might be useful to consider these drivers.

Driver 1: The inherent complexity of nature and society

Problems of nature and society are inherently complex. Addressing them involves aiming to understand, for example, how the brain stores memories, how ecosystems can be resilient to change, how the immune system coordinates a defense reaction to a bacterial infection, or how fragmented terrorist networks function. Complexity compels us to explain the dynamical interactions between system components, which over time give rise to emergent properties that are absent from these components themselves, such as global warming or consciousness. As the National Academies' report put it, a complex system such as climate change cannot be understood comprehensively 'without considering the influence of the oceans, rivers, sea ice, atmospheric constituents, solar radiation, transport processes, land use, land cover, and other anthropogenic practices and feedback mechanisms that link this "system of subsystems" across scales of space and time' (National Academy of Sciences, Engineering and Medicine, 2005). Therefore, understanding complex processes depends on collecting insights from multiple disciplines. Both deep knowledge from monodisciplinary perspectives and expertise in interdisciplinary integration are essential to addressing fundamental questions and problems pertaining to our reality. Collaboration across the natural sciences, the social sciences, and the humanities is crucial to answer such questions more fully.

Driver 2: The exploration of basic research problems at the interface of disciplines

In the previous section, we noted how shared research projects might reveal an increasing overlap between two disciplines regarding their topics of research and their methods. Indeed, initiatives for more interdisciplinary research are triggered by not only societal concern but also scientific curiosity and the desire to explore problems and questions located at the interface of disciplines (Darden & Maull, 1977). It might be the case, for example, that a phenomenon that is the main object of a particular discipline plays a secondary role in another discipline: genetics is only of secondary relevance to the behavioral or brain sciences. One discipline might study the physical nature of a process, while the function of this process might be another discipline's focus: meteorologists are interested in cloud formation, while climate scientists are worried about their contribution to the greenhouse effect, for example.

The borderlands between disciplines have over time become inhabited by new fields such as optogenetics, sociobiology, and digital humanities. They allow researchers to address questions that were previously difficult to answer and enables them to pursue new topics. The exploration of interfaces between disciplines also challenges the existing know-how of researchers, stimulating them to invite researchers from adjacent fields to collaborate and share their knowledge.

Driver 3: The need to solve societal problems

Today, more than ever, scientific research is expected to help solve important societal problems. This expectation has arisen because science and technology provide many techniques and tools that help to cope with major current and future challenges in areas such as food, water, health, energy, and globalization, and more recently the COVID-19 pandemic. Science is often seen as a 'troubleshooter' when it comes to solving complex problems. However, any intervention in such complex phenomena might lead to unexpected changes in a complex system. Indeed, there is usually a bi-directional relationship between technological fixes and societal problems, fixes that often create new problems in addition to offering solutions. Demonstrations of a bi-directional relationship include the impact of artificial fertilizers on water pollution, the increase of automotive transportation and the subsequent rise in highway deaths, the development of life-supporting technologies, pressing concerns around euthanasia, and the growth of urban areas and the related loss of rural land and natural habitats.

Many proposed solutions to societal problems are unsuccessful because they fail to take societal stakeholders into account. Instead of focusing solely on a technological fix, researchers should consider individual and collective responses to proposed solutions that may have an unexpected impact. A well-known problem in medicine, for example, is the lack of patient compliance with a prescribed therapy or medication without the support or supervision of nurses and medics. In another example, the unexpected consequences of the invention of energy-saving lightbulbs included their use for illuminating gardens instead of saving energy. Psychologists and social scientists could have contributed to interdisciplinary solutions that are more 'socially robust' by adding insights into human behavioral responses to proposed solutions and avoiding undesirable effects. Transdisciplinary research methods offer a strategy for such problems, which are discussed below along with an explanation of so-called wicked problems.

Driver 4: Generative technologies

Generative technologies allow for new applications of great value and can transform existing disciplines and generate new fields. They can give birth to new possibilities, thereby widening the scope of both disciplinary and interdisciplinary research. Furthermore, generative technologies such as AI-supported book collection digitization, magnetic resonance imaging (MRI), and the technologies developed at CERN often prompt collaborations between academics from different disciplines and other professionals that would not have occurred otherwise. The widening scope of interdisciplinary research as a direct consequence of newly introduced technologies is also apparent in the recent revolution in data collection and data mining. Increased computational power allows scientists to analyze complex systems in radically new ways. Computer models also allow scientists to understand the underlying forces, interactions, and non-linearities that constitute complex phenomena. The study of the human brain is another example: it dates back thousands of years, but it is only recently that scientists have begun mapping all neurons and their connections as well as the neural and cognitive processes these connections support. Humanities scholars in turn have been able to delve into large collections of online texts and describe how new terminologies developed over time, how collective memory of past events has functioned, and whether historical events were foreseen by a larger public. Projects dependent on these novel generative technologies have also contributed to the development of new research methods and tools, theories and models, and practices and policies. Related interdisciplinary fields have emerged as well, such as computational neurobiology and digital humanities.

Obviously, the four drivers listed above do not cover all the factors that contribute to the increasing prominence of interdisciplinary research. It might be useful to reflect on these factors, as they may push interdisciplinarity in directions that need further exploration. Fuller describes the older 'military-industrial route to interdisciplinarity', for example, which amounts to a trajectory characterized by aims that do not always contribute to societal problem-solving, to say the least. As this trajectory prefigures more common contemporary knowledge production in which universities collaborate with the state (including its military) and industry or commercial firms, he insists that we should consider the loss of academic sovereignty in such forms of interdisciplinary collaboration (Fuller, 2017). In other words, in parallel with the growing role of interdisciplinarity in knowledge production, interdisciplinarians should consider the ethical considerations of their involvement precisely because such projects might have an impact on our societies and environment. It is perhaps not surprising that after World War II, the ethics of science became an area that was more readily embraced, given that scientists during the war were found to have engaged in terrible practices (Resnik, 1998). Since then, scientists have been expected to consider the potential 'dual use' of their results in undesirable applications and to uphold a 'precautionary principle' in their work. Given the very nature of interdisciplinary projects, the researchers involved should indeed make ethical considerations an integral part of their work. This holds in particular for the kind of problems we will consider in the next section.

4.5 Complexity and complex adaptive systems

The first driver we mentioned above – complexity – cuts across disciplines and is widely recognized as one of the main themes in science today. As physicist Stephen Hawking said 20 years ago near the turn of the millennium: "This century is the century of complexity, and complexity and its associated technologies and theories of artificial life, agent-based models, self-organization and the science of networks will revolutionize the way science is done." Problems and phenomena that require an interdisciplinary approach often exhibit the characteristics of complex systems. However, the definition of complexity is a topic of considerable debate. There are some who study complex systems and their dynamics in a mathematical way, while others understand and use the notion of complexity in a much broader sense. In the social sciences, for example, complexity describes the inherent uncertainty of large-scale societal challenges (such as 'wicked problems', defined below) that elude definitive formulations or solutions. As wicked problems are a subset of complex problems, we first focus on so-called complex adaptive systems.

Four basic factors are essential for a system to be classified as complex. The first necessary condition is the presence of a collection of diverse elements or 'agents'. The elements might be atoms, molecules, cells, football hooligans, or multinationals, depending on the complex system being studied. Second, these elements must be interconnected, and their behavior and actions must be interdependent. In other words, they must form a network. The third key criterion is for the elements to display a tendency towards self-organization as a result of feedback and feed-forward loops as well as other interactions that occur on different timescales. Fourth, the elements or the connections between them must be adaptable to change in a local or global environment or must be able to learn (Holland, 2014). The brain is an example of a complex adaptive system (CAS). Its billions of neurons are connected in many ways, but some neural networks might be activated in response to specific stimuli with lasting effects, given the plasticity of neural connections according to the Hebbian principle: 'neurons that fire together, wire together'. For example, after thousands of hours of practice, an expert's brain performs certain tasks differently compared to a novice's brain due to its changing structural and functional properties. Due to the diversity in the connections and interactions between their individual elements, complex adaptive systems often behave in non-linear ways, which means that an output is not necessarily proportional to inputs. A shift may occur from one state to another, or the system can suddenly collapse. For instance, evidence is mounting that global warming may have altered the weather system to a new stable state, making it almost impossible to avoid the climate crisis anymore. Complex adaptive systems do not always change in a linear sense in response to external influences or internal changes, making it sometimes difficult to assess the system's stability. Instead, they have a buffer capacity, making them resilient to change and thereby robust. Oceans were found to buffer a significant amount of CO_2 and global heating, which due to increasing temperatures is now being fed back into the weather system, further contributing to non-linear climate change.

Another key feature of a CAS is the existence of emergent phenomena observable at a more global, macro level but not predictable from – or reducible to – properties of the system's constituent elements at a lower, micro level. These emergent properties arise through self-organizing local interactions, feedback, and feed-forward loops. This phenomenon contrasts with the classical reductionist perspective, which holds that nature is best understood by reducing or decomposing its processes into elementary building blocks that can in turn be analyzed in isolation from other components. Taking the human brain once again as an example: neurons are merely cells passing on signals to one another, yet an interconnected network of billions of cells gives rise to the brain's emergent property of consciousness. A city is another example, with the organization of its infrastructure, countless daily flows of goods, and functions of people living their daily lives even in the absence of a central planner (Page, 2010). The investigation of separate single neurons or humans in isolation would not have yielded results from which we could have derived the existence of consciousness or the nature of city life.

A recent innovative meta-analysis by a group of colleagues from different departments of the University of Amsterdam on the correlation between urbanicity and common mental disorders such as depression and anxiety provides another striking example. Instead of only considering patients, their neural networks, or neurons as elements in a CAS, the analysis also included clusters of symptoms (Conditions A, B, or C) associated with these disorders and the characteristics of urban environments as CAS elements or 'agents'. Consequently, explanations of how such disorders developed revealed that urban factors such as city size, green spaces, and economic development have an impact at different levels, influencing both physical elements such as houses as well as the groups and individuals that make up a city. Interactions and feedback relations between all of these elements on multiple timescales also have an impact on the symptoms that an individual might suffer, depicted in Figure 8 on the next page. Once sets of symptoms co-occur in a frequent and stable manner during certain periods, for example, the symptom network associated with an individual patient risks shifting from a healthy to an unhealthy state, implying that a set of symptoms associated with a certain disorder no longer disappears easily following minor remedies such as a good night's sleep or a supportive conversation (van der Wal et al., 2021). Each element of this extensive multi-level network lends itself to certain interventions, and if enough are addressed in a concerted fashion, the system might return to a healthy, stable state. Many disciplines have produced research that helped build this extensive network, and most of them can potentially contribute to therapeutic interventions.

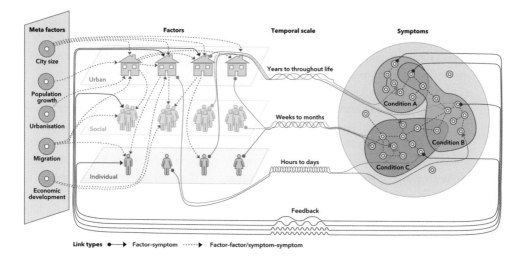

Figure 8   *A complex adaptive network explaining the emergence of symptoms of common mental disorders as a result of complex, non-linear interactions of multiple factors on several timescales. We can distinguish between factors at the individual, social, and urban levels. The characteristics of the urban environment influencing these factors are listed as 'meta factors' in the gray box on the left. (Taken from Van der Wal et al., 2021).*

It is important to note here that a system can be complicated without being complex. Take, for example, a wristwatch. It is a complicated system consisting of many cogs, wheels, and springs, but it is not complex because its behavior is linear – i.e., it is incapable of adjusting or learning, and the connections between its elements remain the same, making its behavior predictable. As a result of its simplicity, a watch is also not very robust. If we remove a single element, it may stop functioning entirely, as it is incapable of learning or adapting or adjusting itself. In addition, a watch does not show emergent properties once it is put together: movements of its pointers, shafts, and cogs can be described with physics that apply to its parts in isolation from each other, and no properties emerge from their interactions that are completely unexpected. A corporation, on the other hand, is complex. When someone decides to retire from a job, the corporation will conceivably adapt to this change, for example by hiring and training someone else. In other words, it will not stop functioning. Because it exhibits robustness, we can say that a corporation is complex.

Complexity is important because we are living in complex times. Thus, complexity is an inescapable reality. The world has become more complex, mostly because of the increased interconnectedness between people at the social, economic, and political levels and at multiple timescales. In contrast, some 10,000 years ago, people lived in groups relatively isolated from each other in different geographical regions and depended only on their own communities. In today's world, however, we have global markets that connect companies and consumers all over the world. Global markets are both highly interdependent and adaptable to change and thus can be classified as complex. In addition, this increased social complexity has had repercussions for the physical and biological world. As a result of global interconnectedness, biological species can spread rapidly over the planet. In addition, problems such as climate change and biodiversity loss are affecting the environment on a global scale. Scientific research aided by new generative technologies – discussed above – have made us even more aware of the complexity of the physical and social world. Furthermore, whereas complicated problems can often be adequately understood by separate disciplines, complex problems need to be addressed through interdisciplinarity because they are multi-causal and multi-layered.

4.6 Wicked problems

While complexity may appear to lend itself to a quantifiable approach, with less room for the kinds of qualitative and hermeneutic contributions that the humanities and the social sciences provide, in actual fact, these academic branches can also be included. Indeed, they may be indispensable, given that the status or function of some elements in a complex system often reflects the interpretations that humans or groups attach to them. For example, loneliness in an urban context can depend on the individualistic or collectivistic nature of one's culture, and this is something that should be reflected in the properties of a complex system, for example in the number of connections and the weights assigned to the connections between its elements – here urban citizens. Yet if ambiguity or multi-interpretability sometimes play a role in grappling with complexity, it is at the heart of another class of problems – so-called wicked problems. Wicked problems are problems that cannot be definitively formulated or solved and are complex by nature. Thus, they are the opposite of tame problems that can generally be tackled using approaches based on a single discipline. To understand and solve wicked problems, additional considerations and multiple perspectives are needed. Examples include climate change, poverty, the COVID-19 pandemic, social injustice, and healthcare. In the latter case, healthcare in wealthy and democratic countries is increasingly difficult to manage, and the costs have become astronomical largely due to a graying population, expensive medications and treatments, lack of personnel, and unmanageable budgets. Therefore, solutions are never simple and depend partly on what society considers to be the minimal conditions for a healthy life. But this raises even more questions. Does viability – as in the ability to live – include being able to work? Should plastic surgeries be paid for by insurance, or only when restoring functions to disfigured body parts? Should we introduce robots during shortages of human nurses, or perhaps only in the care for demented patients who seem to respond well to robot pets? Is there a cap on what we are prepared to pay for healthcare? Clearly, each definition of this specific problem is associated with a particular solution. The concrete tasks of the disciplines involved in responding to this problem are dependent on its definition and its preferred solution, which are a matter of political and social decision-making.

Healthcare is an example of a complex problem with many interacting elements that is at the same time a wicked problem, with normative and cultural multi-interpretability playing a role and therefore without an easy solution to which all parties involved might readily agree upon. A wicked problem such as this one appears unsolvable. Such problems share nine defining characteristics according to Rittel & Webber's classic description (1973), each of which contributes to their wicked nature, which makes them hard to grasp and hence almost impossible to solve:

1. There is no definitive description for a wicked problem. Because of the complex nature of wicked problems, it is impossible to give an exact description. Too many factors underlying the problem influence each other. As a result, the formulation of a wicked problem is in fact a problem in itself!
2. Because solving the problem is more or less the same as defining it, there is no stopping rule – i.e., there is no way to say that you are done. There is no definitive solution, just as there is no definitive description.
3. Solutions to wicked problems are neither true nor false. They are rather defined in terms of better or worse. For instance, a solution might be qualified as 'satisfying'.
4. There is no immediate or ultimate test of a solution to a wicked problem.
5. Every solution to a wicked problem is a 'one-shot operation'. Since every attempt to solve a wicked problem affects the underlying system, it inevitably alters the initial problem. Trial-and-error is not possible: in fact, every trial counts.
6. There is no enumerable set of possible solutions to a wicked problem; moreover, there is no complete overview of permissible or inadmissible solutions.
7. Each wicked problem is essentially unique, and there are no definite classes of wicked problems.
8. Every wicked problem can be considered a symptom of another problem.
9. There is no single way to explain the nature of a wicked problem: numerous possible explanations exist. Thus, the 'chosen' explanation defines the direction of the resolution.

In sum: every possible definition of a wicked problem ends up foregrounding one potential solution while downplaying others. In addition, there is no single optimal solution, as every solution only adjusts *some* features of the problem in a desirable direction and not others – and consequently may be preferred by some parties and rejected by others. At the same time, it is impossible to experiment with multiple solutions, as they always impact on the problem situation itself. This picture may be discouraging to any scientist hoping to contribute to solving the important problems of our time. However, the lesson to be drawn from this challenging situation is that we may have to reconsider the nature of our role and contribution as scientists and to engage with such problems in different ways than we have traditionally done. In the next two sections, we briefly present two methods of research that expand the limits of 'boundary crossing' inherent in interdisciplinarity: transdisciplinary research and action research.

4.7 Transdisciplinary research

In the first section of this chapter, we distinguished between multidisciplinarity, interdisciplinarity, and transdisciplinarity – the latter referring to a mode of research that incorporates not just different disciplinary contributions but also the insights of extra-academic stakeholders into a project. We mentioned that transdisciplinary research can help our research gain more 'social robustness', which means that its results maintain their validity also when implemented as a solution to a societal problem instead of being rejected (or simply not being complied with) by citizens, institutions, or organizations. Now that we have briefly touched upon complex and wicked problems, it is time to reveal how transdisciplinary research works and what makes it particularly suitable for tackling these kinds of problems. For this, we now turn to Figure 9.

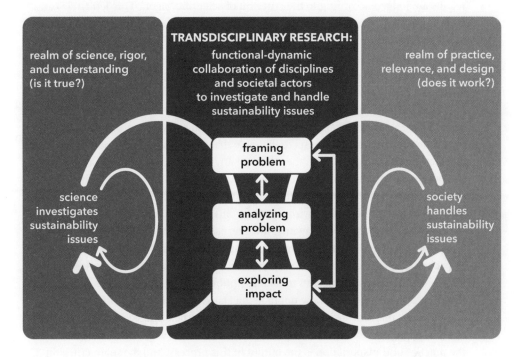

Figure 9 The transdisciplinary research process integrates contributions not only from the realm of science but also from the realm of practice in all phases of the research process (Taken from Pohl et al., 2017). Reprinted with permission from the author.

The middle section of this figure offers a simple representation of the three stages of a research project aimed at producing an impactful solution to a problem. As usual, a research project starts with the phase of framing the problem or definition in such a way that it permits further study, including the formulation of the research question. This is similar to the first two steps that our interdisciplinary research model identifies as the orientation phase. The second stage in this model of transdisciplinary research involves the analysis of the problem, which means gathering insights into the problem and analyzing it. The third stage entails exploring what impact the analysis or insights might have on the problem as originally framed and defined. As is the case with interdisciplinary research, these stages are not followed linearly but may entail an iteration, as when new insights force us to redefine the problem and thus revisit previous research stages.

What is noteworthy about this figure and specific to transdisciplinarity is the integration of two distinct but related processes: the left circle representing the process that occurs within the '*realm of science*' and the right circle referring to the '*realm of practice*'. In other words, if a research project involves scientists whose sole aim is to increase their understanding of a particular problem (e.g., the 'sustainability issues' mentioned in this figure), then the process only moves along the left circle, with the research stages being defined in scientific terms. The aim of this science-only research process is accordingly formulated ('Is it true?'), and this process and its results are judged according to the criterion of scientific 'rigor'. This left circle can then be seen as the science cycle, which we first presented in Chapter 2 (see Figure 3).

Just as science deals with certain problems according to its own methods, aims, and criteria, so too does the 'realm of practice' utilize an approach unique to itself. Taking as an example the problem of sustainability, citizens, governments, and other institutions nowadays must decide on 'sustainability issues' on a frequent basis. These actors generally have a specific 'stake' in such issues, bringing their own values, norms, and interests to the table that might have an impact on how they consider these issues and what they think of any potential solutions. In addition, they also have experiential knowledge with the problem at stake. Each citizen, for example, has some experience with transportation and the efforts, time, and costs involved in navigating their environment for business and leisure. They weigh their public transport options against the alternatives but also consider the costs and reliability of each option. Moreover, using public transport may also have an impact on one's social status, as in many countries only poor citizens make use of this service. If governments aim to increase public transport use in these countries, they may also need to somehow upgrade the status of public transport. Given these complexities, relevant societal stakeholders must also be engaged in the process of knowledge production to answer the question: 'Does it work?'. The design of the practice of transportation is the output of this process, and the main criterion by which that design is judged will be its relevance: does the process tackle the issues that were defined and analyzed as being relevant? And does the design that emerges from that knowledge production process lead to impactful practices? More specifically, will the targeted citizen groups increase their use of public transport, and will that lead to the envisioned sustainability targets?

In many cases of practical problem solving, societal actors can rely upon available relevant scientific data, recommendations, and reports. Other cases may require the production of knowledge that is necessary for the design of societal practices: transportation data, societal attitudes to different modes of transport, local histories of government interventions and their impact, and the potential room in politics and society for new interventions. Problems like these are wicked, as discussed in the previous section. We presented several characteristics that made such complex problems wicked: there is no definitive solution, just as there is no definitive description; since every attempt to solve a wicked problem affects the underlying system, it inevitably alters the initial problem; there is no single way to explain the nature of a wicked problem, since numerous possible explanations exist. The nature of the transdisciplinary research process is tailored to such problems.

Since societal stakeholders are usually part of the complex problem, and because any intervention proposed affects them – with their responses to the intervention further affecting the problem situation – it is essential to include them as equal partners in the research process. Doing so enhances the probability that the research process will succeed in comprehensively addressing the specific nature of the problem, including its normative, socio-cultural, historical, and political features. Needless to say, the involvement of relevant disciplines should mirror the inclusion of relevant stakeholders and their input. According to Pohl and Hirsch Hadorn (2007), there are four principles for designing transdisciplinary research: 1) reduce complexity by both specifying the need for knowledge and identifying the stakeholders involved; 2) achieve effectiveness by attending to the context-specific nature of the problem and its potential solutions; 3) achieve integration through open and balanced collaborations with all involved parties and their perspectives; 4) enhance process reflexivity and be prepared to revisit previous decisions and assumptions in light of newer insights and experiences.

The scope of this book does not allow for us to further expand on these principles or to guide you more elaborately through the transdisciplinary research process. Nonetheless, we do hope that the above gives you some grasp of the overlaps and differences between interdisciplinarity and transdisciplinarity. As universities have also increasingly begun to consider their societal impact as an important goal, this expansion of interdisciplinary research is also gaining traction. This has yet to be the case to the same extent for another mode of research to which we now turn in the closing section of this chapter: action research.

4.8 Action research

Transdisciplinary research aims to balance the production of knowledge and the impactful practices in which knowledge is implemented. The emergence of this mode of research reflects the shift towards more impactful research mentioned above. This approach is one that complements scientific knowledge with instrumental, ethical, and aesthetic forms of knowledge in a communicative process of 'context-specific negotiation' that includes stakeholders (Klein, 2004). While transdisciplinarity transgresses the boundaries maintained by disciplinary and interdisciplinary research, the realms of science and practice are still kept more or less separate even though their interdependence is acknowledged and fostered by the transdisciplinary research method.

Going one step further, 'action research' foregrounds the goal of 'human flourishing' and is hence crucially oriented towards the values relevant for that aim and the actors affected. It is more open to various practices and the ways of knowing that contribute to such practices. Reason and Bradbury define action research as 'a participatory process concerned with developing practical knowing in the pursuit of worthwhile human purposes. It seeks to bring together action and reflection, theory and practice, in participation with others, in the pursuit of practical solutions to issues of pressing concern to people, and more generally the flourishing of individual persons and their communities' (2008). Action research entails that scientists and societal stakeholders establish a democratic process in which everyone participates, jointly developing knowledge and practices that contribute to their well-being.

Action research rose in prominence in the 1970s together with the movements to bring about a societal and political transformation in what were at the time labeled 'Third World' countries. Revolutionary changes did indeed emerge from action research in Latin America, where researchers conducted their work in the service of socio-political organizations or unions that were facing opposition from several sides (Fals Borda, 1979). A major source of inspiration for action researchers has been Paolo Freire's 'Pedagogy of the Oppressed' from 1968, which admonishes investigators to not only research the status quo but also pay attention to how people think about that status quo. Freire argued that research should also support the people in rethinking the problem situation as a necessary step towards changing it. What emerges from this is a mode of research whose results are also judged in terms of their 'actionability', as these should respond to the needs and desires of the stakeholders involved. The results should offer both reliable insights into the problem situation as well as insights that help change the situation according to the needs and desires of stakeholders. Intriguingly, with actionability presenting an additional evaluation criterion for research, action research at the same time relies upon the riches of the imagination as a source of potential solutions and alternative possibilities to the problem situation at hand (Keestra, 2019). In practice, this might lead to the integration of poetry, public theater, music performance, or filmmaking into community-based projects in which action researchers participate. If we were to depict action research in Figure 7 alongside multidisciplinary, interdisciplinary, and transdisciplinary research, it would be seen adding further strands to the integrated results. In so doing, it emphasizes how pluralism is also involved in interdisciplinary integration, to which we will turn in the next chapter

Although interdisciplinary and transdisciplinary research are gradually being recognized as useful modes of research in addition to traditional disciplinary research and are sometimes included – albeit somewhat hesitantly – in many educational and research programs, action research has not found a similar position in most universities and research organizations. However, viewed from a critical perspective, action research is a valuable addition to those other modes of research that are closer to disciplinary knowledge production. We might ask ourselves whether knowledge production should be made more sustainable, which might imply raising questions about the audience for which it is produced, the many forms of costs and benefits involved in it, and so on (Frodeman, 2014). We invite our readers to take some time to reflect on and discuss such critical and methodological questions before embarking on the fascinating and challenging interdisciplinary research process as it is laid out in the second part of this book.

5 Interdisciplinary integration

Before we present our interdisciplinary integration toolbox, let us return to the widely used definition of interdisciplinary research we discussed above:

> *Interdisciplinary research is a mode of research in which an individual scientist or a team of scientists integrates information, data, techniques, tools, perspectives, concepts, and/or theories from two or more disciplines or bodies of specialized knowledge, with the objective to advance fundamental understanding or to solve problems whose solutions are beyond the scope of a single discipline or area of research practice.*
> (National Academy of Sciences, Engineering, and Medicine, 2005)

This definition includes an essential feature of interdisciplinary research: integration. More specifically, it identifies what it is that can be integrated in such research: 'information, data, techniques, tools, perspectives, concepts, and/or theories'. These research elements – which differ quite significantly from each other – roughly refer to different phases of the research project, each of which might lend itself to some form of integration. Below, we briefly explain how integration applies to these different research elements, following the logic of the science cycle discussed in Chapter 2 (reproduced on page 72). You may remember that the science cycle begins with *theories and laws*, from which scientists deduce *predictions*. Testing these predictions yields *results*, which when evaluated generate *observations*. Via an induction process, these can subsequently lead to adjusted *theories and laws*. As each disciplinary contribution in some way rests upon the application of this cycle, the integration of several disciplinary contributions can involve any section of this cycle.

Theories and concepts are typically used to carve out a certain research problem and to determine a specific research question. This part of the science cycle corresponds to the orientation and theoretical analysis phases of a research project (according to our interdisciplinary research process model, which we present in Chapter 6). When different disciplines are involved along with their distinct concepts and theories, these may need some level of integration to ensure that a coherent research problem is determined. After all, the same concept may be defined very differently by different disciplines. For example, the concept of wealth is generally understood by economists in strictly financial terms, the assumption being that rational agents aim to maximize such wealth. Social scientists, however, have a much wider set of definitions of wealth at their disposal, while moral philosophers might choose to work with yet another concept of wealth, for example one that describes wealth as an individual's possibility to develop and express her capabilities and talents in a meaningful sense. Integrating such concepts would require defining a more comprehensive concept or reformulating some definitions in such a way that they at least do not contradict each other. For example, financial wealth can be understood as a condition that enables the achievement of happiness instead of the two concepts being opposed to each other. This shows how concepts that at first sight may not appear to match each other could be integrated not just by redefining them but also by developing a more comprehensive perspective of both concepts. It also illustrates that integration is not just a matter of adding perspectives, given that there may not always be an easy match between them available – if at all.

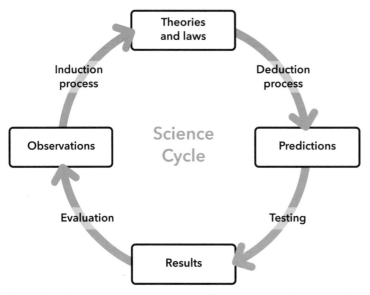

Figure 3   The science cycle, consisting of four processes that connect four components - together providing a (somewhat simplified) representation of science as an ongoing process.

Theories and concepts are typically used to carve out a certain research problem and to determine a specific research question. This part of the science cycle corresponds to the orientation and theoretical analysis phases of a research project (according to our interdisciplinary research process model, which we present in Chapter 6). When different disciplines are involved along with their distinct concepts and theories, these may need some level of integration to ensure that a coherent research problem is determined. After all, the same concept may be defined very differently by different disciplines. For example, the concept of wealth is generally understood by economists in strictly financial terms, the assumption being that rational agents aim to maximize such wealth. Social scientists, however, have a much wider set of definitions of wealth at their disposal, while moral philosophers might choose to work with yet another concept of wealth, for example one that describes wealth as an individual's possibility to develop and express her capabilities and talents in a meaningful sense. Integrating such concepts would require defining a more comprehensive concept or reformulating some definitions in such a way that they at least do not contradict each other. For example, financial wealth can be understood as a condition that enables the achievement of happiness instead of the two concepts being opposed to each other. This shows how concepts that at first sight may not appear to match each other could be integrated not just by redefining them but also by developing a more comprehensive perspective of both concepts. It also illustrates that integration is not just a matter of adding perspectives, given that there may not always be an easy match between them available – if at all.

Another research element that may need to be integrated is *perspectives*, entailing both a theoretical background and a corresponding methodological approach (the latter is discussed in the next section). Perspectives might refer here to different research or disciplinary perspectives on the same problem. For instance, a public health problem might call for an anthropological perspective that sheds light on the influence of culture on the ability of patients to cope with health issues. Alternatively, a demographer might approach the public health problem by looking for correlations between certain group memberships and health conditions: does age matter for a particular condition, for example, or are some religious groups more vulnerable to the spread of infectious diseases? Finally, one could also involve perspectives from different stakeholders of the health problem such as patients, doctors, policymakers, or family members, turning the research project into a transdisciplinary or action research project. Clearly, these perspectives need not contradict each other. On the contrary, they could complement each other and thus contribute to a more comprehensive interdisciplinary perspective on the problem. This could allow for various factors to be considered in parallel, including their potential interactions: group membership and cultural influences might coincide, for example. In addition, perspectives are sometimes understood as levels of abstraction or levels of mechanism. We could, for example, identify a culture or group as the top level of abstraction of the public health problem and then look to the elements making up that group for lower levels of abstraction. Perhaps families and their connections form the next level, with individuals and their interrelations forming another level towards the bottom. Whether or not it makes sense to incorporate even lower levels will depend on the problem at stake: if we are investigating the epidemiology of lung cancer, we might identify a few relevant elements below the level of individuals – such as their lungs and particular cancer-related genes – in addition to higher levels such as families and their genetic patterns or the level of culture with its attitude towards smoking.

Tools and techniques are the research elements that play a key role in the research method and method design. Given the prevalence of pluralisms in science as discussed in Chapter 2, it is unsurprising that methodological pluralism can be encountered within a single discipline, not to mention interdisciplinary research. Researchers apply various tools in their projects, some of which require collaborative efforts to function well. For example, brain imaging techniques – a common research method in cognitive neuroscience – depend on teamwork between engineers, physicists, statisticians, neuroscientists, and cognitive psychologists. The latter might design a particular cognitive task that subjects are asked to perform while their brains are being scanned. A neuroscientist is primarily responsible for determining which activated brain areas or neural networks need to be analyzed, after having co-developed a specific cognitive task with the psychologist. The statistician is responsible for analyzing the huge amount of data, while his choice of statistical tests depends on the research questions, the cognitive tasks that are performed, and the test and control populations. Moreover, the results of brain imaging research can only be interpreted when coupled with other research techniques and tools. Comparative data from animal research, brain lesion studies in live subjects, or postmortems can help to determine the role of certain brain areas in particular cognitive tasks. It is only when these tools and techniques are applied in parallel that the results can be said to be robust – i.e., to hold under a range of different conditions.

Finally, the end results of a research project – *information and data* – can also be integrated. By now, it should be clear that interdisciplinary projects usually yield not just a single set of data or unique insight into a problem. For example, our climate change models are being built from huge amounts of very different sources of information ranging from demographic and socio-economic data to meteorological and oceanological data and results from chemistry research into aerosols. Integrating these results could perhaps mean that researchers aim to determine how the production of certain aerosols in a given region is dependent on the demography and socio-economic development of a certain country or area. They could then proceed to calculate what the contribution of this aerosol production is to the global variation in global warming and ocean temperatures. Clearly, this kind of integration of bodies of information is based on corresponding theories stemming from the different disciplines involved, as these explain the correlation between population size and increased pollution and aerosol production.

In other words, integration is key to interdisciplinarity and can take on different forms while being applied to different phases or elements of the research process. It may not surprise you that, in line with the theoretical and methodological pluralism addressed in Chapter 2, in many cases a research project will employ *integrative pluralism*, whereby multiple forms of integration are applied in parallel to different elements and phases of a single project. This means that integration techniques can differ quite significantly from each other. And these interdisciplinary integration techniques are continuously expanding as well, owing to the development of novel theories, research methods, and instruments along with innovative forms of data processing. For example, due to the development of supercomputers with new forms of data processing, we can now conduct big data science projects that allow us to ask completely new questions or find correlations in huge amounts of datasets that were previously unfathomable. This has led to innovative research questions in the digital humanities, like how the emotions associated with marriage have changed over time, with love and passion only emerging quite recently from the investigation of many digitized historical texts. Similarly, supercomputing is being used as a novel testbed. In antiquity, observations were mostly made *in vivo*, in living organisms. Later, laboratories enabled *in vitro* experiments, making phenomena visible in glass petri dishes or test tubes, for example. Nowadays, researchers can conduct experiments *in silico* by simulating climate change processes with supercomputers, for example. Finally, integration is often a recursive or iterative process and not something that is done only at one single moment, since growing insights might prompt researchers to reconsider previous steps. As Bergmann et al. posit, 'conceptual work and theory building during research in sub-projects require, given a heterogeneous composition of scientific fields, a continuous process of making adjustments, reconciling differences and revising hitherto accepted knowledge claims, since agreement during the process of knowledge integration is something that must always be achieved anew' (2012). Keeping these general observations in mind, we now provide a broad overview of the main categories of integrative tools or techniques.

Keeping these general observations in mind, we will now provide a broad overview of the main categories of integrative tools or techniques to get some grip of the large and expanding collection of these.*

5.1 Main categories of the interdisciplinary integration toolbox

The categories of the integration toolbox presented below differ significantly from each other, ranging from theoretical integration to the integration of the members of a research team. The categories are difficult to compare, which confirms our message that you may not need to choose between them. Indeed, it may often be the case that you will need to apply multiple forms of interdisciplinary integration in parallel: from conceptual integration to the integration of perspectives in a brain imaging study, from exploring the similarities and differences involved in the '*ship of state*' metaphor to arranging a dialogue between extra-academic stakeholders and researchers in order to develop a socially relevant research question. As a result of this disparity, the toolbox of integration techniques contains various sets of tools

---

* This chapter draws on several toolboxes that are available online and in print. The most relevant of these are: Methods for transdisciplinary research: A primer for practice, (Bergmann et al., 2012), with a blog by its main author: https://i2insights.org/2017/05/09/transdisciplinary-integration-methods/; Integration and Implementation Science tools https://i2s.anu.edu.au/resources/tools/; Science of Team Science tools https://www.teamsciencetoolkit.cancer.gov/Public/searchAdvResult.aspx?st=a&sid=1; Wageningen RU Multi-stakeholder partnerships tools and methods http://www.mspguide.org/tools-and-methods; td-net methods for co-producing knowledge https://naturalsciences.ch/co-producing-knowledge-explained together with ETH Zurich TD lab, https://tdlab.usys.ethz.ch/toolbox.html; Toolbox dialogue initiative https://tdi.msu.edu/research-overview/tdi-integration-research/; Toolkits for Transdisciplinarity OEKOM Verlag https://www.oekom.de/publikationen/zeitschriften-gaia/toolkits-for-transdisciplinarity/c-168. More recently, the global Inter- and Transdisciplinary Alliance is developing an inventory of such toolkits at https://itd-alliance.org/inventory-project/.

that range widely, from dialogue techniques to visualization techniques, from computational models to policy advice, and so on. Below, we outline the main categories of integration methods in our toolbox.

5.1.1 Theoretical and conceptual integration

The starting and end points of the science cycle are the theories, which consist of concepts as well as the principles, ideas, or laws that determine the relations between those concepts. Theories and concepts are linguistic elements that all scholars use when they want to pinpoint precisely and unambiguously a particular research object and one or more of its properties or behaviors. This is a challenging task, as in many cases scientists use words that are also used in everyday language – which is not particularly known for its unambiguity, precision, and specificity but which is still the medium of extra-academic, experiential knowledge. In our aim to be more precise and specific, we may find that the everyday use of a concept lumps together objects and phenomena that do not in fact share crucial properties with each other. One example of this is the practice in Germanic languages to refer to whales as fish, even though they are in fact mammals. Conversely, we might also find that the use of certain concepts has the effect of 'splitting up' a set of objects or phenomena in such a way that it defies understanding. For example, psychologists still disagree on whether intelligence is a single phenomenon or whether emotional, social, and cognitive intelligence are fundamentally distinct from each other.

Even more generally, does it make sense to combine research on the basis of conceptual similarity – for example, research into human consciousness, the collective consciousness of a group, and animal consciousness? Should we lump these together or rather split them up, not letting ourselves be guided (or perhaps misguided) by common language? To get a grip on a complex concept, we do need to make use of further concepts, which inevitably implies that we are *establishing relations between several concepts* and the phenomena or objects to which they refer. These linkages offer new venues of investigation that often bring different disciplines together. Such conceptual relations can be further pursued as *'heuristics'* or strategies for further research, as they open up areas of research that can clarify certain aspects of a complex phenomenon like consciousness (Keestra & Cowley, 2009). Bringing together disciplines as different as entomology, animal ethology, neurology, psychology, and history, scholars can begin to specify when they consider their research objects to be sentient or awake and when they are not, thereby perhaps discovering general processes that they share. Integrating lines of disciplinary research into phenomena that are conceptually related, researchers have in fact established an increasingly refined conceptual framework or taxonomy of consciousness. Using conceptual distinctions and connections, scholars from different fields are bringing their research to bear on shared objects or phenomena and working together towards further conceptual refinement. This may require them to 'lump together' or 'split up' these sets of objects or phenomena in order to increase the consistency in their use of concepts – for example by developing a conceptual framework like the one referred to here.

But how does the discipline of history fit into this conceptual framework? Can we talk about the 'consciousness of a nation'? While insects (entomology) and humans share not only an evolutionary history but also many neurobiological properties and response patterns, nations don't seem to fit in here. Hence it seems plausible that consciousness might in some sense apply to both insects and humans while nations are distinct in this regard. This brings us to another form of conceptual integration, a form that has been productively employed for ages by a wide range of scholars: the metaphor. A metaphor involves transferring a meaning to a target object that is only in a limited sense comparable to the source object. However, by articulating comparable properties between the two, insights about the source object's properties can be used to hypothesize about similar properties in the target object. For example, a nation needs in some sense to be alert to changes in its environment and to exert a form of conscious control in order to respond adequately to those changes. Applying the distinctions and relations of the concept of 'consciousness' as heuristics to the object 'nation', historians and political theorists are invited to determine analogies to the different levels of consciousness in nations or to discover where sentience occurs in the socio-political arena of a nation. Perhaps it even makes sense to refer to some nations as being almost comatose and others as fully alert and to establish some such taxonomy for nations? Instigating the investigation of such parallels, metaphors bring together disciplinary insights that initially might seem difficult to integrate.*

Having elaborated on conceptual integration, it is now obvious that developing an interdisciplinary or transdisciplinary theoretical framework offers ample opportunity to integrate contributions from various disciplines. As theories allow us to explain and predict the properties or behavior of relevant objects (or phenomena), insights from different disciplines about these properties can be integrated into an increasingly comprehensive theory. Taking consciousness once more as an example, sociologists and social psychologists have found that the contents of someone's consciousness also depend on societal values and norms. These have an impact on how people attend to features and objects in their environment and in so doing determine what is being cognitively processed by individuals. Consequently, sociologists might contribute some concepts and principles – about the role of specific values in particular groups or societies, for example – to a theory explaining and predicting what the contents of consciousness might be for members of those groups versus those of out-group individuals.

---

* Since this brief look into the toolbox of integrative techniques is inevitably inexhaustive, we would mention here the use of 'boundary' concepts or objects as integrative tools. These concepts or objects are shared across the boundaries separating different research fields or disciplines, allowing both sides of the boundary to make use of them – often because of their relatively loose or ambiguous meaning. A visual image of an object or a report about that object might play such a role, with interpretations on both sides diverging from each other, making the integration of their interpretations a valuable interdisciplinary goal.

Finally, it is important to note that, as we found to be the case with concepts, scholarly progress is not only made by theoretically bringing together different sets of objects but also by splitting up a set of objects by developing different and more specific theories for subsets of them. An example of the latter is when Einstein's theory of general relativity determined that Newtonian mechanics only holds under certain conditions – more specifically, the theory stated that when objects are extremely fast or heavy, Newtonian space-time no longer applies to them. As a result, Newton's theory no longer has the general applicability it was previously thought to have had but instead is more circumscribed, which has been demonstrated to be extremely important in cosmology and quantum physics in particular.

5.1.2 Integration of research methods and instruments

As explained earlier, science is partly characterized by the use of special methods and instruments replacing direct sensory perception of bits of our world. These methods and instruments can help to detect phenomena that might otherwise escape our perception, for example by using telescopes or large corpora of texts. Or they allow scientists to have numerous identical observations of a certain phenomenon by producing it repeatedly in a lab experiment under controlled circumstances. Or they force scholars to demonstrate the soundness of their interpretation of a certain legal principle by aligning it with their reading of a great number of similar statements in other relevant texts. In other words, scientific methods help us to yield specific results that are acceptable across individuals and across different situations – results that are not immediately available in everyday life. No wonder, then, that disciplinary differences between methods may also create obstacles or barriers between researchers. Methodological differences imply not only practical differences but also potentially conflicting assumptions about what adequate methods are, what holds as an adequate proof, what kinds of data should be produced, and so on (Lélé & Norgaard, 2005). Since these assumptions are often implicitly held by a 'community of scholars' within a discipline, integrating these multiple methods typically requires the articulation of their underlying assumptions.

However, we also mentioned earlier that even single disciplines are in many cases already characterized by methodological (in addition to theoretical) pluralism. By employing different research methods, scientists can yield more robust results than when they investigate a phenomenon only according to a single method, given that each method has its own limitations and vulnerabilities. Using different methods in parallel or in a sequential order reduces the risk of such flaws, as one method's strength might compensate for another method's limitation. Moreover, some methods even rely on each other. For example, collecting a statistically reliable set of data on voters' opinions requires the use of an adequate poll or survey, a common quantitative method in the social and behavioral sciences. The list of questions or statements that voters need to respond to is itself often based on a series of in-depth interviews that researchers hold with a representative number of voters. These interviews are carefully scrutinized and compared in order to find relevant differences and similarities between voters that can in turn be compared with each other. Consequently, the results of a small-scale qualitative interview method can be integrated with a quantitative survey method in order to tap the opinions of a large group of people.

This example of sequential (step-by-step) integration demonstrates how methodological integration is not uncommon even within a single discipline. Still, when researchers of different disciplines engage in an entirely new and interdisciplinary project, they may need to be creative and develop their own integrated set of methods tailored to their shared research project. Clearly, they must first agree upon a conceptual and theoretical framework as well as the research questions, since the methods for answering those questions must be determined. When agreement is reached, the researchers should check whether the methods that participating disciplines contribute separately can meet the requirements set by the research questions. It may well be the case that the gaps in knowledge emerging in an interdisciplinary project allow for investigation along disciplinary methodological lines when a particular discipline's methodology is 'borrowed' by another discipline (Klein, 1990, see section 4.3 above). For example, investigating the influence of additional social factors on voters' opinions might merely imply the expansion of the existing survey instrument to a novel domain for their integration, yet it does not require a wholesale change of the methods.

A more systematic integration of methods may be appropriate when entirely different fields of research or levels of explanation need to be combined. This integration itself can take several forms. One discipline's research method might be expanded or adjusted such that it can be applied to the domain of another field. The application of quantitative and statistical methods was initially prevalent in economics and some fields of psychology and was only introduced later to sociology. As these methods are designed to determine and analyze correlations and other patterns in large numbers of data, they require social scientists to adjust their traditional focus on their domain. For example, this integration implied a decreasing role for specific contextual knowledge or individuals' considerations in the social sciences (Calhoun & Rhoten, 2010). Adopting an adjusted version of a research method from another discipline implies that the source discipline is prepared to reconsider what it means to collect data.

A more comprehensive integration of methods and instruments takes place when those of two disciplines are successfully combined. For example, if we would like to better understand – and perhaps even predict – the cognitive and neural processes involved in voters' decision-making, cognitive neuroscientific methods and social scientific methods must be integrated. There may be several options available for tracking these micro-level processes in cognitive neuroscience, for example by using fMRI brain imaging methods, eye-tracking methods, or EEG brain-scalp measurements. Investigating the meso-level behavioral outcomes of these processes might be accomplished by using a survey instrument or by letting subjects play computer games, for example. Having theoretically investigated what cognitive mechanism might explain voting behavior, researchers subsequently need to assess which methods and instruments are able to spatially and temporally track the cognitive and neural processes underlying this mechanism. Integration might then imply presenting subjects whose brains are being scanned with social scientific survey questions or computer games while correlating subjects' voting behavior with dynamic changes of brain activation patterns (cf. Vander Valk, 2012).

A very different integration of disciplinary methods occurs when models are involved, particularly computational or simulation models. Such models are considered integrative interdisciplinary objects and therefore will be discussed in the next section.

5.1.3 Integrative models and objects

In many cases, scientists have limited access to their research objects or can only manipulate them in very limited ways. Consequently, many disciplines have developed or discovered an object that could partly replace their research objects, using models of those instead. A well-known example is the traditional planetarium that astronomers crafted to investigate planetary motions, for example, in which case a model is meant to represent the real cosmos. Obviously, such a representation is highly abstracted and reduced, and it is important to explicitly articulate what has been changed compared to the source object.

Somewhat different are the application of animal models for research of human physiological, behavioral, or cognitive phenomena. In such cases, the appropriate animal must be chosen to satisfactorily represent the research object, which means adequately determining the relevant processes involved and making sure that these can be investigated in the animal. For example, nematode *C. elegans* can represent many relevant genetic processes found in humans, while chimpanzees and bonobos are assumed to represent human social processes in relevant ways. Determining such a reliable animal model can require input from anthropologists, ethologists, animal physiologists, neurobiologists, psychologists, social scientists, and perhaps still other scholars, as they must consider whether or not the constraints and limitations of the target model rule out its being representative for the particular properties of the source object under scrutiny. Social scientists sometimes determine a model population that represents the 'average community' under study. Similarly, historians or cultural analysts might define a certain collection of items to be adequately representative of the relevant phenomena under investigation, as when the combination of a country's progressive and conservative papers are selected to faithfully represent a country's psychological mood during a certain period.

An increasing number of disciplines do not employ such reduced or limited models but instead develop their own computational or simulation models in order to retrodict the statistical patterns found in existing datasets or to predict patterns that might emerge from subsequent empirical research. Such models allow for the explicit integration of certain factors or variables, which need to be weighted and connected to the other variables of the model. This leads to the elegant integration of a discipline's research interest and data into another discipline's model, namely in the form of an additional factor or variable. In developing a computational model of historical language evolution and making use of huge data from digitized book collections, for example, the model could be adjusted to include the geographical distribution of the individual books, creating a computational model that now allows for the emergence of local variations in this model.

With animal models or the partial integration of computational models already being helpful, a more intense collaboration occurs when the integration of specific research domains into a more comprehensive system is facilitated by way of the integration of computational models representing several of its subsystems (Jakeman, Letcher, & Norton, 2006). Current computational climate change models are developed via such integration, with their large numbers revealing that the strengths and limitations of each single model might force scholars to make use of more than just a single such integrative model (Le Treut et al., 2007; see Figure 10 in section 5.1.5). Averaging the results of the outcomes produced by a number of models does offer some guarantee that the biases or bugs of a single model will not skew the overall modeling outcomes, as it is not to be expected that these biases or bugs across models are caused by identical algorithmic peculiarities.

Comparable to such use of computational models, researchers from different disciplines may also include a set of different variables in a so-called optimizing or optimization function. Such a function can be employed to find an optimal solution from a large set of possible solutions that the function can produce as a result of tweaking the variables. Such optimization functions are used, for example, when policymakers aim to develop an optimal mobility strategy for their region that considers changing demographics, the expected number of car drives, economic growth, types of employment, and so on. Sustainability policies often also rely on the outcomes of optimization functions, which include variables and factors such as taxation, consumer behavior, pollution effects, and animal loading. As much as an optimization function can help to develop and assess different scenarios, they remain simplifications of the real world, just as models are. Indeed, interdisciplinary scrutiny of such optimization functions may help to render them more robust by integrating additional variables that were previously neglected.

Models are not only employed as a tool for integrating contributions from different disciplines and providing another 'test bed' for producing different scenarios and outcomes. Manipulating the variables that determine a model's behavior also allows for its use in explaining why test data show certain patterns. This kind of explanatory integration will be discussed further below in section 5.1.5.

5.1.4 Integration of data and results

You may recall that the definition of interdisciplinarity cited at the beginning of this chapter also referred to the integration of information and data. This might sound strange, as if researchers could simply add one set of data to another, even though they might be very different. However, these differences might be complementary to each other in some sense, providing for a more comprehensive or robust data collection. Hermeneutic research of corpora of both theological and legal texts might enable scholars to compare interpretations of divine and human forms of 'responsibility' and 'causality', and this comparison might inform conceptualization in both disciplines. It could lead to the formation of a more elaborate conceptual network than each single discipline might have yielded, allowing for more nuanced and robust interpretations in both theology and law.

In most cases, different sets of data or results are produced using different methods, with different theoretical frameworks supporting their production. Notwithstanding these differences in background, integrating different sets of data is certainly possible. Similar to the integration of subsystem models into a more comprehensive model, one set of data might be complemented by another collection. This complementarity might apply to their being harvested by means of different methods, even though they are meant somehow to represent the same phenomenon. We observed this in the interdisciplinary research of voting behavior which used different disciplinary methods. If voting behavior results are confirmed by the verbal responses of voters in interviews, the robustness of voting results and our interpretation of these results increases accordingly. However, it is not uncommon for sets of data to show systematic differences even though they are meant to represent the same phenomenon. This requires an explanation: why do subjects vote differently from what their verbally declared position is?

As much as different sets of data might complement each other as their flaws or biases are non-overlapping, other forms of complementarity can be even more productive. Representing distinct factors, two or more sets of data can enable researchers to detect more complex and nuanced patterns in the combined datasets. Assuming that all datasets are similarly constrained, their comparison or combination might be possible. The integration of datasets of annual national gross domestic products with results of demographic studies might lead to insights into the impact of demographic changes or ageing populations on economic growth. Such an integration might require the adjustment of datasets in such a way that their integration is not contradicted by their representing different levels of scale – as when we must average local demographic data in order to be integrated with domestic product numbers, which might only be available at the national scale. Having discussed concepts, methods, models, and data, we now turn to explanation, which is in many cases the overall goal of scientific research.

5.1.5 Integration via explanation

Researchers are most satisfied if they can not only predict the outcome of an experiment but also explain this phenomenon. Questions about when, why, or how a certain phenomenon occurs require an explanation. An explanation can hold for a specific case, or it can be applicable to a large set of cases provided they share relevant properties and there are no influences that might affect some cases but not others. We mentioned earlier that, in contrast to what many scientists and philosophers of science have held for some time, explanation in the form of general laws is more the exception than the rule. In the life and social sciences, we might be able to explain a phenomenon due to a causal or dependency relation, yet this relation might not behave in a lawlike manner, as many intervening factors could create a rather irregular pattern. Because of this, we should accept that interdisciplinary projects yield a variety of kinds of explanations. Perhaps the most crucial distinction in types of explanation is that between full and partial explanation (Ruben, 1992).

Starting from this distinction between full and partial explanation, it follows that interdisciplinary research might contribute to developing a more comprehensive explanation, as the perspective of additional disciplines might help to add explanatory factors. Especially when different factors interact with each other, for example through a feedback or feedforward loop between them, a partial explanation will be seriously flawed if it is effective at all. Take, for example, the explanation of climate change: over time it has become ever more comprehensive by including ever more factors or subsystems in the climate system. This is clearly illustrated in Figure 10 below, which represents the increasing complexity of climate models from the mid-1970s to the appearance of the AR4 model of the Intergovernmental Panel on Climate Change in 2007 (Le Treut et al., 2007). Initially, the explanatory model only involved acid rain, carbon dioxide emissions by industry, and solar radiation. When geoscientists and oceanologists were able to produce sufficient insight into the interactions between climate change and processes like erosion or heat absorption by oceans, for example, these were included in the explanatory model. Not only were the scientists able to add to the general explanation of climate change, their contributions could now also help to explain some variation in the available data. For example, oceans can absorb and release heat in certain patterns over time, which did help to explain why the overall increase in global temperature was less than expected for a period of time and greater than expected sometime later (Heffernan, 2010). In other words, by adding a factor, the explanation was not only more comprehensive; with gained insight into the interactions between the factors, but it was also more precise and robust, leading to a better understanding of the phenomenon.

As complex as the explanatory model AR4 of the 2007 IPCC report is, the represented factors are largely physical and chemical in nature. However, the presence of ships on the ocean and industry in the countryside already indicates the presence of a human factor which, when included, necessitates an explanatory model to integrate factors such as politics, psychology, and sociology, which are very

The World in Global Climate Models

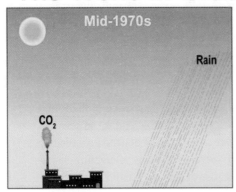
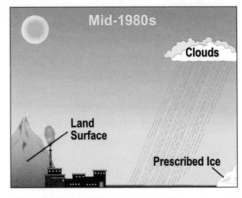
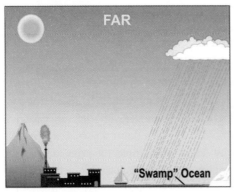
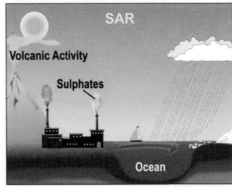
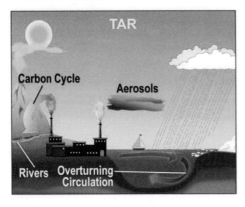
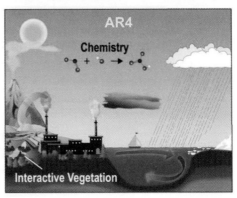

Figure 10 *The evolution of climate change models as developed by the IPCC, becoming increasingly complex and offering higher temporal and spatial resolution over time (Le Treut, Somerville, Cubasch., Ding, Mauritzen et al., 2007). Reprinted with permission from the author.*

the model must accommodate for the influences of ambiguous language, historical events, the impact of powerful individuals and groups, and so on. Clearly, this makes such an explanatory model much more complex and thus affects how we can use it. While it might have been possible to retrodict and predict climate changes with the different from the physical factors. With those human, behavioral, and social factors, use of a computational model matching this explanatory model, with the addition of these human factors, we may be forced to develop a set of scenarios that capture multiple potential trajectories of the system that are all plausible yet variously influenced by these factors.

The same applies to the mechanistic explanations that are typically produced in the life sciences and increasingly so in the social sciences (Bechtel & Richardson, 1993; Hedström & Swedberg, 1998). In cognitive neuroscience, for example, research yields an explanation in terms of components that interact in an organized way and together produce a certain cognitive phenomenon like consciousness, visual perception, or the understanding of action. Although many events in the human central nervous system and brain can be explained in electrophysiological and neurochemical terms, we may also need to include effective environmental factors in our mechanistic explanation. Understanding human behavior may require us to include socio-cultural, political, religious, or other influences (Keestra, 2012). To begin with, someone's responses to a certain image or situation might depend upon her understanding of it: the swastika symbol has very different meanings depending on whether it is in a European or a Hindu context. Norms and values have an impact on our perceptions: some subjects will perceive cattle as holy, others will consider pigs not fit for consumption, while still others will perhaps respond to meat eating in general with disgust. Factoring in such influences yields an explanation that not only contains a description of the brain activities involved in disgust responses but also why this happens sometimes but not in other instances – for example, when no taboo is at stake.

Explanation of human behavior in humanities disciplines like history, theology, or anthropology might entail the articulation of the motives and reasons that an individual or a group has for their verbal or physical behavior. Why did Arjuna team up with Krishna to engage in battle? Why did the French Revolution turn violent and not produce the general emancipation and liberty that many were hoping for? Explaining individual and collective behavior, scholars will turn to relevant texts and interpret the beliefs, intentions, expectations, moods, and other ingredients that are relevant to their protagonists against a wider socio-cultural and historical background. Obviously, such explanations may also include more quantitative factors – as, for example, when income disparity in 18th-century France might help to interpret intentions and moods.* Insight into the widespread poverty among the French population at the time might also lead social historians or economists to conclude that a revolution was perhaps not optimal in meeting its economic needs.

---

* As the examples used in this section show, Dilthey's distinction that is often made between causal explanation and understanding – or between *Verstehen* and *Erklären* – is not upheld here. That distinction relies very much on the notion that explanation should always be in lawlike, mathematical terms, and since we do not ascribe to that restricted notion, the distinction is of limited use to this book.

Indeed, when an interdisciplinary investigation yields such a comprehensive explanation, the scientists involved could make use of it when developing an appropriate practical intervention or instrument.

5.1.6 Integration in an intervention or instrument

Interdisciplinary projects do not always or exclusively result in new knowledge. Transdisciplinary projects in particular tend to also produce practical interventions or instruments. Still another mode of research – action research – is developed in such a way that it produces 'actionable knowledge', with its actionability being one of the criteria by which the research project can be assessed (Keestra, 2019). The final phase of transdisciplinary projects is often referred to as the 'transformation' or 'implementation' phase, usually entailing the production of a social or political intervention, a technical instrument, an informative exhibition, or another medium that brings about change (Hirsch Hadorn et al., 2008). A medical research project, for example, is in many cases dedicated to the development of a form of treatment, a therapy, a prevention measure, or a medicine, or to the adjustment and improvement of any of these things. New knowledge must of course be included in such cases, yet it is put to the service of developing or improving such practical interventions. The integration of insights from different disciplines should at least partly benefit these practical interventions, making them more effective or robust in different situations, for example.

In many cases, creating change in the real world by implementing a certain strategy does indeed require input from a large number of disciplines, as so many factors might affect the desired change. In an Alpine forest protection project, for example, a group of researchers and relevant stakeholders developed an integrated action plan consisting of numerous measures directed at forestry, agriculture, wildlife and nature, and finally also public relations – all of which had to be coordinated and consistent with each other (Hindenlang, Heeb, & Roux, 2008). Not included in this list are several constraints that these measures are likely to have been subjected to, like the project's financial resources and the personnel needed for implementing the measures. For selecting an optimal set of measures within these constraints, the research group will most likely have used optimization functions to accompany their scenario-building and comparison.

Although largely absent in this book, let us not forget to mention the engineers, architects, designers, robot builders, and others working in interdisciplinary teams to develop their 'boundary objects' – objects shared by several disciplines or fields. In many cases, they start from an already existing object on which they work together to improve, enlarge, apply in novel contexts, or otherwise modify. The nature of such a collaboration has its impact on all phases of the research project, as the research question is then formulated in terms of technical requirements, mechanical options, or other practical targets. Reaching those targets will require taking adequately the material and mechanical properties, ergonomic design for end users, economic cost-effectiveness, and other factors into account. Projects like these are often conducted by following the stages of 'design thinking', which includes elements that are not always prominent in interdisciplinary projects such as creative thinking, developing, and testing prototypes. Nonetheless, projects like these are interdisciplinary or transdisciplinary in that multiple disciplines are involved along with additional input from practically oriented experts and relevant stakeholders.

Expanding a project team to include still more kinds of collaborators and stakeholders makes the use of collaborative tools even more important than usual. Let us therefore end this chapter on integration by briefly examining integration methods that are dedicated to the functioning of the team itself.

5.1.7 Integration of the research team and its members

The interdisciplinary integration toolbox presented in the previous sections consists of methods that help to integrate those elements that are somehow part of the knowledge production process. Following the definition of interdisciplinary research quoted earlier, these methods are applicable for integrating concepts and theories, research methods, models and objects, data and results, explanation, and interventions or instruments. In a way, then, we can say that these methods are content-focused rather than process-focused. This section focuses on the fact that in many cases, interdisciplinary research is a process involving a team and not just conducted by individual scientists. Given the obstacles that an interdisciplinary and transdisciplinary research team must overcome before they are even able to apply such integration methods, we should pay some attention to the challenge of creating an integrated team of various experts and stakeholders. As we will discuss collaboration elsewhere in this handbook more extensively, this section will be relatively brief.

Clearly, developing interdisciplinary integrated insights usually requires a team – leaving aside exceptional, neo-Renaissance, multitalented individuals – consisting of members with different expertise and backgrounds who might also be implicitly holding different norms and assumptions regarding the research project and its value. Unsurprisingly, given these differences, the team needs to safeguard an adequate level of integration as a team if it is to succeed in producing relevant and integrated results. For example, without sufficient individual and team reflection and communication, even the best set of individuals will fail to produce integrated knowledge – as many historical examples of failed projects show (Keestra, 2017; O'Rourke et al., 2019). To be optimally productive and creative with each other in the face of a complex problem, the integration of the team must be taken seriously.

Supporting this observation, it is of interest to note the recent emergence of a new scientific discipline called 'Science of Team Science' initiated by a section of the National Institutes of Health. Focusing partly on evidence-based insights about effectively collaborating teams of scientists, this discipline is gathering a growing number of items for their toolkit which consists of four sets of methods focusing on managing the team, managing the research, managing the project, and disseminating the results to a wider audience.* Managing the team requires that the team use certain methods to improve communication and to avoid bias or miscommunication among team members, for example. Management tools abound to help the project management and planning involved, while decision-making tools might be required to constrain the large number of options that are available when a team is dealing with uncharted territories. Dissemination of team results to non-scientific stakeholders can be facilitated by the use of specific visualization tools, as these can help to make the results understandable to a lay audience.

Similar to the development of the Science of Team Science is the argument made for a specific discipline of 'Integration and Implementation Sciences'. Since most interdisciplinary and transdisciplinary research projects aim to solve complex real-world problems that include various unknowns by bringing together scientists and stakeholders with various backgrounds, it is argued that there is a need for specialists to facilitate and support such projects (Bammer, 2013). These integration and implementation specialists have expertise about relevant techniques and tools for integration like the ones presented in this chapter. In addition, they can support the team in managing unknowns and making decisions about whether a particular gap in knowledge must be filled or not, whether it is a matter of translation difficulties between knowledge bases, and so on. Finally, with their expertise in dealing with obstacles to evidence-based policies or the cultural factors that influence societal practices, they can assist the team in finding optimal ways to bring the project results to fruition (Bammer et al., 2020).

Underlining the importance of structured reflection and communication between team members is the prevalence of dialogue methods in this context. There are numerous dialogue methods, each of which have their own focus. Some help the team to gain a broad understanding of the problem at stake by integrating different perspectives of scientists and stakeholders, considering differences in insights, goals, and values. Others are more focused on a shared discussion of normative issues, interests, perspectives, and worldviews (Bammer, McDonald & Deane, 2009). More specific still is the Toolbox dialogue method, which was developed to support a research team in articulating and discussing differences about the fundamental assumptions that individual members hold regarding knowledge and science, the value and application of research, norms and goals that should guide research, and so on (Looney et al., 2014).

In sum, although it is important to carefully consider what integration techniques can be applied to bring together different disciplinary contributions, we should not forget that it is people who are involved in doing this, which makes it essential to spend time and attention to the integration of the members of the team.

---

* See https://www.teamsciencetoolkit.cancer.gov/Public/searchAdvResult.aspx?st=a&sid=1 for this toolkit.

5.2 Final remarks on integration

Since integration has been hailed as the crucial element of interdisciplinary and transdisciplinary research, it is not surprising that we have spent some time on it. Before moving on to the next part of this book, it is important to add a few closing remarks on the use of these instruments. First, it will be apparent by now that many projects will require the application of more than just a single form of integration during more than just a single project phase. Second, any single discipline might contribute to more than just a single mode of integration. Third, it is not uncommon for a form of 'cross-integration' to take place. Cross-integration occurs when the results stemming from a disciplinary investigation of a project's sub-question have an impact on the theory of another involved discipline, or when the question asked by a group of stakeholders is answered by the use of a method of a specific discipline, enhanced by the instrument engineered by another. Fourth, the list of categories and examples mentioned here is by no means exhaustive. On the contrary, in many cases a team will need to develop its own package of integration tools and methods, making it a unique experience that utilizes the team's joint creativity and imagination.

Finally, we would be remiss if we did not offer some words of caution. Although integration is such an important element of interdisciplinary research, we should not close our eyes to the possibility that different perspectives on a given complex problem do not lend themselves to seamless integration. Recall our discussion in the previous chapter on so-called wicked problems, the definition of which is itself a matter of debate affecting potential research questions. We used the example of healthcare in affluent societies to show how societal and political choices about finances, the robotization of care, the availability of plastic surgery, etc. would determine the disciplinary research tasks and their integration. Each choice will result in a specific prioritization – of finances, say, over objections against robotization or over the availability of plastic surgery for aesthetic purposes – which has an impact on the role that disciplinary insights can play. There is no single optimal solution to such a problem – a reflection of its wicked nature. So even though this book emphasizes the importance of interdisciplinary integration, we should always be careful not to assume that all perspectives can equally be integrated into the more comprehensive solution that interdisciplinarity has to offer. Clearly, the challenge remains to argue why a particular integrated result should be preferred in any given case. Thus, doing interdisciplinary research requires those involved to reflect, make explicit, and argue with each other more than in other forms of research, creating extra challenges but – we hope – also more opportunities for mutual learning and inspiration.

Part 2
The Manual – 'The How'

In Part 2, we guide you through the interdisciplinary research process. We begin by pointing out the different phases of this process, and then we describe it in its entirety and indicate which parts might be of particular interest to interdisciplinary researchers. More specifically, we will walk you through the problem and the research question, the theoretical framework, the methods and techniques of research, the data collection and analysis, and finally, the discussion and conclusion. As integration is a defining characteristic of interdisciplinary research, we will illustrate the different integration techniques introduced in Chapter 5 using examples of peer-reviewed research.

The steps you need to take to complete an interdisciplinary research project might overwhelm you. As an undergraduate, for example, you might not feel comfortable dissecting the theoretical background of your own discipline. But do not be discouraged: you will find that after going through the process a couple of times, you will increasingly feel at ease being an interdisciplinary researcher. All beginnings are hard, to some degree. But when you have the right mindset and attitude, you will succeed!

6 The interdisciplinary research process

Although there are many similarities between the disciplinary and interdisciplinary research process, interdisciplinarity will often have additional questions to answer, steps to perform, and challenges to overcome. Compared to disciplinary research, it is considerably more focused on the integration of different methods and/or theories and is therefore often a more iterative decision-making process that requires much more collaborative action. We have developed a model for performing interdisciplinary research, represented below in Figure 11. Building on this model, this chapter describes the different phases of the interdisciplinary research process, focusing on the additional challenges that are specific to interdisciplinarity.

6.1 The interdisciplinary research model

Figure 11 depicts our model of interdisciplinary research. It describes a generalized research process in which the different steps are distinguished for each research phase. It is important to note that there is not one standard research process, not only because research processes differ in practice but also because what is considered a normal research process differs from discipline to discipline. Therefore, the model we propose here should serve merely as a guideline during your own research, not as a strict protocol.

In Chapter 11, we run through an example of a complete interdisciplinary research project using this model as a guideline in order to help explain how this research model can be operationalized.

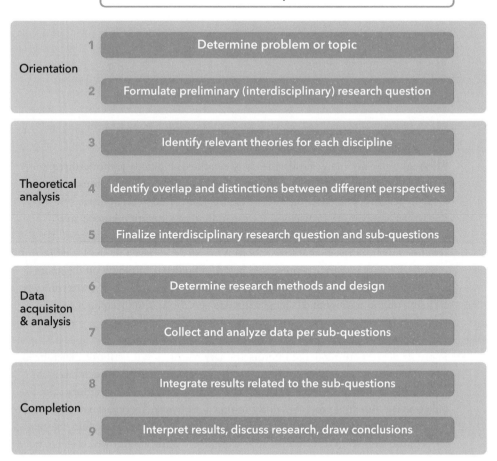

Figure 11   The interdisciplinary research model consists of four main research phases. In each phase, different tasks are performed to realize a well-integrated research project.
NB: As research is often an iterative process, it might be necessary to return to a previous step before moving on to a new phase of your research process.

While expanding your knowledge on your topic, you may sometimes need to return to a previous step, given that research is often an iterative process (see Chapter 4). However, the order of the steps to be taken is more or less fixed, which implies that you should not skip any step. An obvious example is that you cannot analyze data that you have not yet collected. However, it is important to realize that you must think ahead (i.e., you need to know how you are going to analyze your data before you start collecting them). For this reason, we have grouped together several steps in the following phases of the interdisciplinary research process.

Integration

In this interdisciplinary research model, we distinguish the separate steps (indicated in the blue text boxes in Figure 11) that must be completed in each particular research phase (yellow text boxes). Although for some tasks, integration may not seem to be the most obvious approach, in many interdisciplinary projects it is necessary to achieve integration at multiple steps. For example, the theoretical framework will probably integrate elements from theories from different disciplines, or your methods might consist of different disciplinary methods.

Remember that in conducting interdisciplinary research, you are entering a field that is possibly completely new to you and perhaps even to your supervisor. As you will be connecting disciplines and fields, you will be unable to rely on your prior disciplinary experience to guide you. Even when your research is focused on a particular case study, you might often find that you can rely only in a limited sense on preceding studies. For this reason, you will need to set out a detailed, clear-cut process to guide you as you try to engage in the holistic thinking required for interdisciplinary integration. This model of interdisciplinary research will serve as a general guideline, but you must map out in detail how you will apply its guidelines to your specific project.

Phase 1   Orientation

You might start your research process by choosing a topic that fits your interests or a problem that you would like to solve. In all cases, you need to explore the topic and find out whether the topic is suited to an interdisciplinary approach. It is important that you identify whether the disciplinary expertise within the research team is relevant for solving the (preliminary) research question and how the different team members may complement each other. This entire process is what can be called the orientation phase.

One of the biggest challenges of interdisciplinary research is ensuring that each contributing discipline has its relevance within the research project and that no single discipline becomes too dominant. By formulating a preliminary research question, you define the focus of your research. This helps to set the boundaries for your research project (see Chapter 7). While this may feel like a constraint, it often makes your project more realistic and achievable. Moreover, this forms the basis for the creation of a theoretical framework, which you will develop in the next phase.

Phase 2   Theoretical analysis

As explained in Chapter 2, what distinguishes scientific knowledge from other knowledge is the role of theories and laws, which are typically absent from experiential and more traditional forms of knowledge. It is for this reason that phase 2, with its focus on theoretical analysis, is very important. Preparing scientific research requires the development of a theoretical framework, which is usually drawn from a literature review. Such a review involves an overview of the most relevant and up-to-date theories on the research topic, a systematic analysis of the most important insights, and the identification of any so-called knowledge gaps in this field. Further, in the case of interdisciplinary research, this theoretical analysis will not only survey the relevant publications from the various disciplines but also pinpoint the differences and commonalities between the disciplines.

When analyzing the different disciplinary parts of the theoretical framework, you should be constantly aware of the different disciplinary points of view with regard to concepts, theories, methods, and type of results. For example, the definitions of certain concepts can differ between disciplines (e.g., evolution in biology versus astronomy). What also may differ between disciplinary traditions is whether certain methods (e.g., quantitative and qualitative techniques) are qualified equally important.

What is unique to interdisciplinary research is that often this increased awareness of different disciplinary perspectives will enable you to ask an insightful integrated interdisciplinary research question (see also Chapter 5).

As a last step, it is time to consider the best way to answer your research question. What are the sub-questions you need to address to fully answer the main research question, and which disciplines can handle these questions? Please note that sub-questions can be both interdisciplinary and monodisciplinary.

Note: It is important to realize that the research process is an iterative process, i.e., the cycle of operations is executed more than once, gradually bringing the project closer to some optimal condition or goal. In interdisciplinary projects, this iterative process is even more important, as full integration is often not reached after one cycle. So instead of being a simple linear process, you might need to revisit a previously made decision in light of a decision you are making later in the process.

Phase 3   Data collection and analysis

Before data acquisition and analysis takes place, scientists often first consider the most suitable methods for answering the sub-questions. But as integration can also take place at the methodological level in interdisciplinary research – e.g., using a genetic algorithm (biology) on datasets for disease diagnosis (medical sciences) – it is useful to describe in your theoretical framework the proven effectiveness of this method in previous studies.

When you have completed the theoretical stage of your research and considered the research method (steps 1 to 6), you should now have the information necessary to write a research proposal. It may appear strange that devising a research proposal would consume so much time and attention but writing a research proposal is a crucial part of your research. You must have a well-thought-out plan and a good indication of the kind of results you will acquire in order to answer your main research question. So, the more effort you put into writing a well-integrated research proposal, the more likely it is that your research project will be a success.

After your research proposal has been completed and accepted, you can begin the practical stage of your research and start collecting data. Once you have collected the data needed to answer your sub-questions, you must think about ways to analyze them. Different disciplinary perspectives can be integrated into your analysis methods and interpretation of the results. A good example of such an integration is a study performing a meta-analysis on the results of qualitative studies on health-related ICT support for seniors (Vassli & Farshchian, 2008). For this study, expertise in both qualitative as quantitative methods were necessary not only to carry out the analyses but also to interpret the outcome.

Phase 4 Completion

As presenting research results always entails a consideration of its limitations and future extensions or follow-ups, the project is completed by integrating all the sub-questions and providing a conclusion and discussion. Formulating a conclusion and discussion in interdisciplinary research might be more difficult than in monodisciplinary research because in this phase you must integrate the results and insights related to your sub-questions. You must then ensure that you actually answer or solve your main research question and draw the most important conclusions of your research project by referring back to your integrated theoretical framework. The differences and overlaps you have found in phase 2 (theoretical analysis) between the different disciplinary insights will help you understand the implications of your results and will give you more insight into the topic. The discussion of results in interdisciplinary research projects will also often consist of an explanation of the limitations of previous monodisciplinary research on the same topic.

6.2 Interdisciplinary collaboration in research

An interdisciplinary project is typically a challenging task for a research team, with collaboration playing a crucial role. As you can imagine, an interdisciplinary research project is a major assignment that often requires more 'people and knowledge management than other types of research, as you will be working in a team of investigators from different disciplines (Figure 12).

When a team consisting of members from different disciplines discusses details, it is easy to see how misunderstandings come into existence. What may be fundamental for one scientist may be overlooked or considered unimportant by another who is working from a different angle or discipline. Without in-depth reflection on how interaction takes place and how differences can be overcome, much valuable knowledge might be overlooked or lost. Thus, a greater awareness of the team dynamics may help those involved to overcome obstacles and improve the quality of the teamwork as well as the individual work.

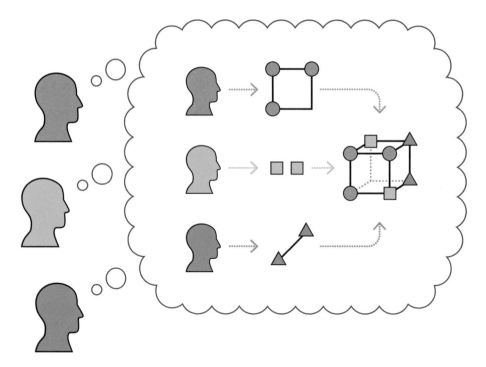

Figure 12 An interdisciplinary team of experts develop together a more comprehensive understanding of a phenomenon – represented by the three-dimensional cube composed of different elements each of them contributes. Team reflection (upon interdisciplinary collaboration) facilitates the process of integration of their possible distinct ideas of the phenomenon (Figure adapted from Keestra, 2017).

The successful use of team diversity as a source of knowledge and innovation requires openness and academic flexibility as well as collegiality. Moreover, given these demands, interdisciplinary team projects tend to require more attention and patience than is often the case in monodisciplinary research. As a result, successful team management is an important contributor to an effective interdisciplinary research project. Therefore, much effort should be invested in: (1) keeping a balance between the contributions of involved team members, (2) coordinating the different sub-projects, and (3) designing the project planning.

6.2.1 Team relations

It is important to realize that in an interdisciplinary team, the involved members may not only differ in their knowledge; personal differences in confidence, motivation, sense of security, and communication style may also arise. To make sure that everyone is equally involved in the project, addressing these personal differences is very important. By entrusting one of the team members with the extra task of making sure that everyone agrees on the way forward and of increasing the team's confidence, potential problems on team relations may be solved. As team conversations on the relations between team members can touch upon sensitive things, an attitude of openness and respect is essential.

6.2.2 Tasks on content

The orientation phase is the phase in which there should be room for every thought, but the interdisciplinary team should work towards a shared understanding of the phenomenon studied (Figure 12). Although disciplinary differences should be appreciated (as these often lead to interesting insights), it is nonetheless important for these differences to be valued strictly in accordance with the overall goal of the project. Often, discussions on content will become intertwined with those about the process or with inter-team relations. As these are separate issues and can lead to misunderstanding and conflict, it helps to make one team member responsible for facilitating discussions about the separate parts of the project (i.e., the research question, the theoretical framework, the methods, etc.). This person should also monitor whether the different sub-questions (which may change over time) remain relevant to answer the main research question.

6.2.3 Team process

The sub-projects are likely to be executed at different moments in time. For example, the results of one sub-project might form the starting point of a second sub-project. Therefore, it is essential to map out a thorough but flexible plan (and system for managing data) when you are about to begin your interdisciplinary research project. Of course, some parts of the project may turn out to be more time-consuming than expected, or they can have implications for parts of the project that you expected to have already completed, forcing you to return to previous steps. Given these complexities, it is particularly important to realize from the get-go that composing and writing your final research report can take a considerable amount of time. Develop a flexible plan that allows you to cope with such unexpected surprises and try to leave some time over at the end of the project to allow you to finish the final report properly. Once again, it helps to assign to individual team members the tasks of chairing meetings, devising a sound project plan, coming up with a data management plan, and – importantly – ensuring the project's execution.

6.2.4 Interdisciplinary communication

To solve the complex problems that interdisciplinary teams are often faced with, communication is extremely important. Experts often distinguish between three categories of teamwork: task, process, and relation, which means that, in line with these aspects, three types of conflict can arise in teams: task conflict, process conflict, and relation conflict. Communication is at the root of many team conflicts and also the key to resolving these. For example, differences in disciplinary language may cause problems in communication (e.g., the term 'vulnerability' can mean something very different to economists and anthropologists), so common ground should be reached. But communication is also necessary to successfully guide the process of team members building on each other's specialist knowledge. Ultimately, the different sub-projects from the different disciplines should fit together to answer the main research question, which will often be a complex process.

Exercises

6.1 Getting to know the team

- In what way do the research steps of an interdisciplinary research project differ from those of a monodisciplinary project?
- Give your top three most important factors for a successful interdisciplinary project and discuss these with another team member.
- When starting an interdisciplinary project, it can be useful to get to know each team member. First produce an overview of your individual characteristics by answering the following questions:

    1. What are my strong characteristics (i.e., what do I bring to our team)?
    2. What are my weaknesses (i.e., what do I need from our team)?
    3. What does the team need to know about me?
    4. What are my project success criteria?

For a related team-exercise you could also perform this exercise:
(scan QR-code below)

After these exercises, evaluate the individual answers and discuss:

1. Whether there are any collective pitfalls (e.g., has everyone time management difficulties)
2. Whether there are any mismatches in the team's expectations
3. How to overcome these potential challenges (e.g., what mutual agreements are needed)

For more in-depth information on collaboration, see:

7 The problem

In any research project, the first task is to narrow down the problem or topic to a research question. Especially in interdisciplinary research, this can be a challenging step, as it often involves a problem studied from multiple perspectives or disciplines. Here, we describe the first two steps of our interdisciplinary research model (Figure 13). The central topic of this chapter is how to narrow down a problem to a preliminary research question that is feasible in terms of both time and scale.

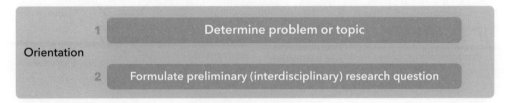

Figure 13 First two steps of the interdisciplinary research model

Considerations:
- Which disciplines are most relevant?
- What are the dominant perspectives of those disciplines regarding the problem or area of interest?

Step 1 Identify problem or topic

Finding a research topic that triggers the academic curiosity of all team members is not always easy. An additional challenge specific to interdisciplinary projects is determining a problem that can be addressed by the available disciplinary expertise in the research team. A triangulation exercise may help you find a suitable topic (see exercise 8.1), as it allows you to visualize the overlap between involved disciplines and thus gives insight into the possible topics that are relevant to most team members.

Once you have decided on a topic or a problem that you want to focus on in your project, you need to set a starting point for your research. The following questions can help you in your initial literature research and may further clarify the context of your topic of research.

1 What do I already know about the problem?
2 What aspects related to the problem are important to consider?
3 What other problems does it relate to?
4 From which perspectives can I look at this problem?

It can be very helpful to visualize your answers to these questions in a mind map or a similar illustration. By drawing such a mind map, you can visualize and connect as many relevant aspects pertaining to the topic as possible. Furthermore, this can help you to decide on a more specific topic of overlap between disciplines, which is necessary for the next step in the process: formulating a preliminary research question. Figure 14 shows a mind map built around the topic of disaster risk reduction.

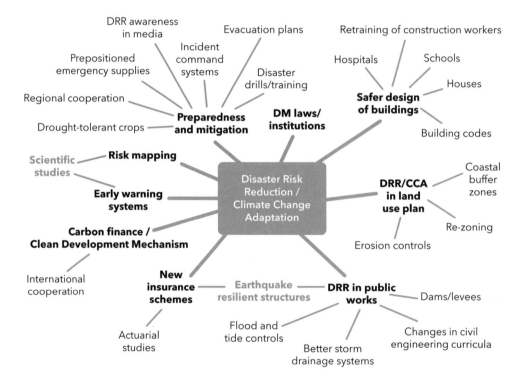

Figure 14 Example of a mind map on the topic of disaster risk reduction (DDR). This mind map gives an overview of different topics and possible perspectives on DRR. In this example the sub theme 'earthquake resilient structures' was connected to two other sub topics on DRR, which could point to an opportunity for a good interdisciplinary research topic (Taken from Talisayon, 2010). Reprinted with permission from the author..

Another way to reveal the most relevant aspects pertaining to the research topic is the development of a mechanistic model. This is often a visual representation of a system that can be investigated by examining its individual parts. It is commonly used in the life sciences and the cognitive sciences but is also gradually being adopted in the social sciences. Discussing a mechanistic model in this step of your project may also help you realize how the next steps in the interdisciplinary research process are connected.

A typical example of a mechanistic model is shown in Chapter 8 (Box 2). The interdisciplinary topic underlying this mechanistic model concerned the direct and indirect consequences of fisheries on ecosystems. By extending an already existing mechanistic model, the research team identified additional topics of integration for better fishery management to better account for the ecosystem functioning of fisheries. Another example of a mechanistic explanation is the climate model shown in Chapter 5 (Figure 10). Over the years, climate scientists have discovered additional components that are part of the mechanism that constitute our climate system, and the figure illustrates the evolution of the explanatory model from the 1970s onwards. While a mechanistic model can be useful at various stages of research, it is mostly valuable as an integration technique for determining a more specific interdisciplinary topic.

Once you have identified the interdisciplinary problem or topic, you need to know which perspectives are best suited to address this problem. The first item of order here is to identify the disciplines that are most relevant to the problem or topic. When you consider all relevant theories and methods around this topic, which disciplines are the most important for researching the problem, and how do you select the most important disciplines for your problem?

A good starting point would be to find an overview of disciplines and their main topics put together. Other particularly helpful sources of information are review articles in which a researcher (or group of researchers) answers a research question or maps the state-of-the-art knowledge regarding a theme or a problem by analyzing and integrating the results of tens or sometimes even hundreds of relevant research articles. These review articles are a good source for getting an up-to-date overview of the perspectives within a field on a particular subject or problem. With the knowledge gained from these review articles, you can further refine your mind map into a more theoretically informed one.

Step 2   Formulate preliminary research question

At this point, you have decided on a topic, identified disciplines relevant to that topic, and created a mind map in which the main disciplinary theoretical perspectives are presented. This will help you to formulate a *preliminary* research question. It is a preliminary research question because you will be refining and adjusting it according to the theoretical and methodological insights that you gather when developing a theoretical framework (the next step in the research process; see Figure 13). However, you do need a preliminary research question, as without a clear focus on what you want to research within your topic of choice, it is easy to get lost in the vast amount of academic literature.

When setting up a preliminary research question, try to avoid specific disciplinary biases, as they may complicate interdisciplinary collaboration and integration further down the road. It would be wise at this early stage to avoid – where possible – unnecessary jargon, technical terms, or even non-technical terms that are characteristically used by academics from one discipline. One way to avoid these disciplinary terms is to try to formulate the problem in everyday language. During subsequent steps of the interdisciplinary process, you will probably re-examine the definition of the problem and develop more precise wording that speaks to all the relevant perspectives (Newell, 2007).

In the end, your research question must meet multiple criteria. But it helps to have these criteria at the forefront of your mind also when working on your preliminary research question. The goal is a finalized research question that is relevant, anchored, researchable, and precise. Below, we specify what we mean by these terms:

Relevant The question should be related to the broader societal or academic problem you wish to address, reflect the reason for your research project, and be the driver of interdisciplinary research. In short, it should be clear why it is worthwhile to seek an answer to the question.

Anchored The question should be the logical outcome of your literature review, expert interviews, and theoretical framework. Your research question must be embedded in the fields of knowledge of your research topic, and the result should be of added value to the fields involved.

Researchable It should be possible to conceive research methods that can address the question in the amount of time and with the means available.

Precise The question should be straightforward and specific. It should be clear what the research focus is.

Examples of research questions that are not reasonably narrowed down according to these criteria include:
- How can we improve sustainable agriculture?
- Is the judiciary affected by criticism?
- What is the best cure for depression?

There are many flaws in these research questions. For one thing, the results from previous research, which are usually much more specific, have not been incorporated into the question (e.g., 'How can we improve sustainable agriculture?'). All three questions contain concepts that lack specificity (e.g., 'judiciary' and 'criticism'). With regard to the third question – 'What is the best cure for depression?' – there are too many possible types of depression for a research project to be able to investigate.

Examples of research questions that are more reasonably narrowed down:
- To what extent can fogponics contribute to sustainable agriculture?
- What are the effects of societal and political criticism on the judiciary in the Netherlands?
- To what extent can selective serotonin reuptake inhibitors (SSRIs) increase the effectiveness of cognitive behavioral therapy in patients diagnosed with depression?

One of the most difficult aspects of this step in the interdisciplinary research process is to incorporate all relevant perspectives (disciplinary or otherwise) in your finalized research question. Try to avoid a one-sided question that leans too much towards one discipline's theory. At the same time, you should make sure that the question is not too broad but instead relates to a specific element of the topic. It is for this reason that interdisciplinary research often requires multiple iterations before a research question is formulated that is relevant, anchored, researchable, precise, and balanced in terms of interdisciplinarity.

Exercises

7.1 Finding a shared topic of interest via 'triangulation'

If an interdisciplinary team consists of a predetermined combination of disciplines, it may be somewhat challenging to develop a shared topic, given that scientists tend to start with a topic and then assemble the necessary team. However, determining the 'overlap' between several disciplines can be done by performing a 'triangulation' exercise.

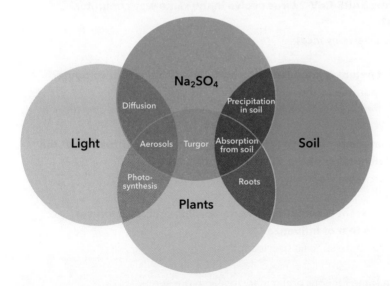

Figure 15 An example of triangulation. Combining insights from chemistry, geography, plant physiology, and physics, a group of students decided to investigate the impact of sodium sulfate from volcanic eruptions on the environment, particularly on plants (Gelauff, Gravemaker, Isarin & Waaijen, 2015). Reprinted with permission from the author.

You might similarly determine a specific and shared topic by following these steps together:

1. Write down individually and from your disciplinary perspective a list of topics that interest you, considering what might be of interest to other disciplines.
2. Exchange the different lists of topics with each other and look for one or more topics that show similarity or overlap with others.
3. Consider individually what sub-questions arising from such topics your discipline could answer and formulate a preliminary research question.
4. Again, discuss in the team the different topics and the sets of sub-questions related to those topics that each member has formulated, choosing the most relevant or fruitful topic for further elaboration.

7.2 **Preliminary research question: relevant, anchored, researchable, precise?**
Start this exercise with the preliminary question you are currently working on in your project. Then, reformulate your preliminary research question into four new variations of the question, trying to cover the *four* criteria for research questions (relevant, anchored, researchable, precise).

For example:

> Preliminary research question
> **Do computer viruses evolve in the same way biological viruses do?**
>
> **Q1: Does the SARS-CoV-2 virus evolve in the same way computer viruses do?**
> (High(er) societal *relevance*)
>
> **Q2: To what extent is knowledge on mutational hotspots in computer viruses comparable to those in biological viruses?**
> (More *anchored* in 'computer virus literature', and it is also more *specific*)
>
> **Q3: Do estimations of the ILOVEYOU computer virus outbreak meet the reproduction number of the COVID-19 outbreak?**
> (Better *researchable*, 'reproduction number' is quantifiable)
>
> **Q4: To what extent is evolution in metamorphic computer viruses comparable to that of influenza?**
> (More *precise*, more specific species/forms)
>
> Now do the same for your preliminary research question!

8 Theoretical framework and research question

Now it is time to turn the preliminary research question into a researchable research question. By developing a theoretical framework that addresses your preliminary research question, you will be able to describe the 'state of the art' in terms of theories and laws in your field(s) of interest, sharpen your ideas, and then – in line with the iterative fashion of research – formulate a better, more specific research question (Figure 16).

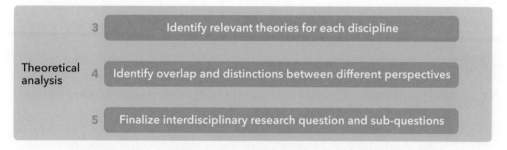

Figure 16 Steps 3, 4, and 5 of our interdisciplinary research model. Developing a theoretical framework to formulate the research questions.

Considerations:
- Consider all relevant theories, concepts, and assumptions that each discipline can contribute.
- Where do disciplinary theories overlap? Where can you find or create common ground?
- Is it possible to integrate the theories of the relevant disciplines into one comprehensive theoretical framework?

As discussed in Chapter 5, communication between (disciplinary) perspectives is a first crucial step towards integrating them. Therefore, an important question is how to enhance communication within your interdisciplinary research group to combine your knowledge into a theoretical framework.

In general, an understanding of your own discipline is needed if a useful conversation between academics from different disciplines is to occur. In other words, you should be familiar with the theories, concepts, methodologies, and associated assumptions that are central to your discipline and realize how these and other ingredients make up the science cycle according to which scientific research of your topic is conducted (see Chapter 2). You could also answer the following questions as a way of bringing the disciplinary perspective to the fore (partly after Paul & Elder, 2012):

- What is your disciplinary focus and perspective on the problem?
- What insights is this perspective based on?
- What are the strengths and weaknesses of this perspective?
- What are the borders of your disciplinary perspective in researching the problem? That is to say: where do you see opportunities to collaborate with other disciplines?

The second and arguably most important skill to enhance integration through communication across disciplines is metacognition (i.e., insight into your own level of knowledge) and reflection on your assumptions (disciplinary or otherwise), mindset, and background (Keestra, 2017). All researchers operate with implicit assumptions that are based on personal values and are guided by their discipline (Lélé and Norgard, 2005). When communicating with other researchers, keep in mind your own values and assumptions, and realize that these are personal and not set in stone. Keep an open mind towards others and their values.

Thirdly, as explained in Chapter 5, a main precondition for interdisciplinary integration is the ability of researchers to explain their diverse backgrounds, their perspectives, their insights into the research problem, and their methods of investigation. In practical terms, this means you must be able to articulate and communicate your perspective on key concepts and assumptions in a way that someone who does not share your specific disciplinary background can comprehend. For example, when explaining your perspective, avoid using jargon or technical terms. It also means that, when a colleague from another discipline is explaining her perspective, you must dare to ask questions. Critical, open, and respectful communication with academics from other disciplines is one way to discover the assumptions underlying a particular discipline.

Step 3 Identifying relevant theories for each discipline

Now that you have a clear overview of the related concepts and assumptions, you can conduct literature research to develop a theoretical framework. Such a framework enables a broader understanding of your research question by framing that question in terms that are included in specific theories. This theoretical framework serves as the 'backbone' of your research question and will help you select the information or knowledge that is most relevant to your research. Since theories are generally built on data gathered by using specific research methods, the theoretical framework will also help you to identify the most useful methods for studying this topic.

Note that, especially in the context of interdisciplinary research, the development of the theoretical framework is an ongoing process: you will be working on the framework throughout the first half of your research project. As your research progresses and you learn more about the topic, you will also need to update your theoretical framework. This means you will be iterating between your research question(s), your theoretical framework, and your methods. Your integrated theoretical framework will ultimately be a substantial part of your research proposal and final report and will help you to further anchor a relevant final research question.

Collecting disciplinary insights into the problem

To gain insight into the state of the art of relevant disciplines, it is important to find out which ideas and theories have already been developed within these different disciplines. Make sure you list the major publications within the fields that are relevant to your research project; a good start is to look for papers that are well-cited, published in high-impact journals, or written by major researchers in the field. You can look through the reference list of relevant publications to backtrack the line of evidence and set alerts on keywords in paper databases to receive the latest research updates in your field.

Another good starting point would be to interview an expert on the topic. An expert can help you to refine your literature search and choose the most relevant perspectives to research a topic.

To begin constructing an integrated theoretical framework, you may need to create several more specific theoretical frameworks from different disciplines before you are able to decide on how to integrate their most relevant parts. The integrated theoretical framework should:

- offer a critical overview of relevant academic literature on your research topic and question;
- be based on perspectives and theories from each of the selected disciplines, preferably from more than one field (or sub-discipline) within each discipline;
- be a coherent story, as opposed to just a collection of concepts or theories, and show the reasoning behind the integrated perspectives that led to your research question.

Table 2   Data management table. This provides an overview of the relevant research fields, theories, concepts, and assumptions for your interdisciplinary research.

Full reference to the book or article				
Disciplines / sub discipline	Theory / hypothesis	Concept(s)	Assumptions / methodology	Insight into problem
Name and describe the specific involved research field(s) and specialization.	Describe the relation between the (f)actors that are considered (e.g. cause and effects) or why a certain intervention is thought to be useful to overcome the research problem.	Give clear definitions of the concepts and explanations in the theoretical framework and explain which of the (potentially plural) definitions you will take as a point of departure in your research project.	Analyze the basic assumptions underlying your theoretical framework and explain which assumptions you will incorporate, or which you reject.	Explain how the theory and the key concepts it entails help to provide more insight in or a possible solution to the problem you are addressing. Take also into consideration possible limitations.

How can you find the relevant research fields, theories, concepts, and assumptions for your interdisciplinary research? Below, we indicate where you might find the various elements of the data management table.

Discipline/sub discipline

To find out which discipline the insights you find useful for your research are coming from, the following may help:
- Check the research field of the journal or the affiliation of the author(s). What kind of institution/organization do they work for?
- Check the mind map you made when working on your preliminary research question. What research fields are linked to your problem?

Theory

It can be difficult to ascertain the theory that lies at the basis of the insights you find useful. Here, we provide some advice on how to do so:
- Sometimes a prominent theory or hypothesis (e.g., the theory of evolution, the Marxist theory, or the Goldbach conjecture) is explicitly named in an article, and it might even be taken up in the abstract. In that case, it is only a matter of looking for a good description of what the theory states.

- When the theory is not explicitly stated in the article, it can usually be found in review articles. These types of articles not only provide the state-of-the-art theories in the research field, but they also often present and compare a variety of possible explanations.
- Another strategy for finding out about what theory the author(s) adhere to is to find more information about the author(s) themselves (see the first suggestion under 'Discipline/sub-discipline').

Concepts

To identify the concepts that are relevant to your research project, the following guidelines are useful:

- Concepts answer the 'what' question: what is the research project about, i.e., what are the phenomena under investigation? Think of keywords, which are also often listed on the first page of journal articles.
- Beware that the 'same' concept can be found in different theories and in the context of different theoretical frameworks, giving different interpretations of such a concept (e.g., rationality, chaos, and equilibrium).
- Concepts can sometimes be difficult to distinguish from theories, principles, causal links, phenomena, or methods. It might help you to realize that a theory is an overarching framework, while a concept is one of the defining key elements of that theory.
- It can be helpful to create a concept map in which you summarize different aspects of this concept.

Assumptions

To find out what assumptions are underlying certain concepts, the following guidelines may help:

- Often the key concepts are defined in the first paragraph in the introduction of a paper.
- When an overview is made of different definitions from a variety of papers/disciplines, it provides an overview of possible different interpretations of this main concept.
- Ask experts from the different disciplines involved about contrasting assumptions they have experienced with regard to your main concept.
- Especially when researchers from more distant disciplines (e.g., chemists and philosophers) are part of an interdisciplinary team, discuss the norms and values that prevail within the research practice of each discipline.

Insight into the problem

Summarize the previous steps, i.e., how each discipline/perspective may contribute to solutions or insights into the main topic. Below are two examples of a data management table filled in with information on disciplinary research on the link between alcohol consumption and aggressive behavior.

Table 3 Example of a data management table

Caetano, R., Schafer, J. Cunradi, C.B. (2001) Alcohol-related intimate partner violence among White, Black and Hispanic couples in the United States. Alcohol Research and Health, 25, 58-65

Disciplines / sub discipline	Theory / hypothesis	Concepts	Assumptions / methodology	Insight into problem
Psychology Understanding behavior and mental processes **Epidemiology** Patterns of health and illness at the population level	**Subculture of violence theory** Certain groups in society accept violence as a means of conflict resolution more than others. **Social structure theory** Socioeconomic factors characterize lives. **Acute effects hypothesis** Alcohol disinhibits aggressive behavior.	**Intimate Partner Violence (IPV)** **Violence** 11 physical violence items from the Conflict Tactics Scale. **Alcohol problem measures** Survey with 29 alcohol-related problems **Problem syndrome** Clustering of problems in one area with other factors	**Higher rate found among female-to-male IPV** Might be due to underreporting or clinical samples. **Coding alcohol consumption** >3 drinks/day is heavy drinking. **Violence** Only reported physical violence, not emotional abuse **Coding ethnicity** No mixed ethnicity category	**Support for subculture of violence theory** Black subjects reported higher f-to-m IPV than white subjects. **Alcohol as excuse** Presence of drinking in IPV does not mean alcohol is the cause. Alcohol problems do not always cause IPV, but can be used as **marker for identifying at risk population**.

Table 4   Example of a data management table

Fish, E.W., Faccidomo, S., Miczek, K.A. (1999) Aggression heightened by alcohol or social instigation in mice: Reduction by the 5-HT B receptor agonist CP-94,253. Psychopharmacology, 146, 391-399				
Disciplines / sub discipline	Theory / hypothesis	Concepts	Assumptions / methodology	Insight into problem
Behavioral neuroscience Physiological and developmental mechanisms of behavior **Neurochemistry** Neurochemicals that influence networks of neural operations **Genetics** Genetic variation, specific genes, and heredity in organisms	**Individual differences in brain chemistry can predict behavior** Psychopathological behavior can be treated clinically with medicine. **Genes affect behavior** Individuals with genetic predisposition to drink alcohol exhibit tendencies towards impulsive violent behavior.	**Receptor and agonist** An agonist is a chemical that binds to a receptor of a cell and triggers a response by the cell. An agonist often mimics the action of a naturally occurring substance.	**Treatment works on curtailing behavior** Pharmacological agents towards serotonin-receptor sub types may have behaviorally specific anti-aggressive effects. **Animal models do not transfer to human trials completely** The receptors are not identical, but function homologous.	**Social context impacts behavior** Social stimuli that precede an aggressive encounter potently modulate arousal. **Biological correlates of behavior** might explain some differences in aggression between individuals. There may be a genetic component for aggressive behavior, elevated by alcohol.

> **Box 1**
> ## Alcohol consumption and aggressive behavior – analysis of the data management table
>
> Suppose that your topic of interest is why alcoholics are often more abusive toward their family members than toward others.
>
> The first article reviews several theories on alcohol-related intimate partner violence among White, Black, and Hispanic partners. Alcohol consumption goes hand in hand with increases in aggressive behavior toward partners. It is unclear, however, whether alcohol should be considered the cause of aggression. Certain expectations, the history of the individuals involved and of their relationship, and environmental factors also play a role when violence occurs.
>
> The authors of the second article discuss social instigation or stimuli that precede the violence, and they address variants in brain receptors that somehow modulate the level of aggressiveness.
>
> At first glance, the different insights all provide a piece of the puzzle that you are trying to solve. But problems could lurk below the surface, such as:
>
> - Do both articles adhere to the same concept of violence?
>   - Does this match your preconceived definition of violence?
> - What level of alcohol consumption are both articles talking about?
>   - How does that compare to the 'alcoholics' you want to study?
> - Do the articles favor the social correlates of violence over the neural correlates, or is it the other way around?
>   - In the first article, the level of analysis of brain biochemistry is not considered to be relevant to the behavior under study. In the second article, however, the social circumstances modulate aggressive arousal in the brain; this modulation is described in biochemical terms, without a mention of individual or social history.
> - Animal studies vs. studies with humans.
>   - The second article does experiments with mice, whereas the first deals with human subjects. Are the 'mouse insights' relevant to understanding the human condition?

Step 4 Identifying overlap and distinctions between different perspectives

Using a data management table will not only enable you to detect harmonious insights from different disciplines in the literature. More importantly, it will help you to discover insights that conflict with each other. These conflicting insights offer an opportunity for interdisciplinary integration in the process of finding common ground. For example, are the apparent contradictions between the results perhaps because the insights are derived from studies with different age groups or in different countries? That might point to a developmental influence or a cultural or social one, which might facilitate a more comprehensive interdisciplinary understanding.

Analysis of supporting and conflicting differences

To find supporting and conflicting insights, you can examine whether the insights around the same topic reveal distinct aspects. Preferably, insights that support each other should stem from different lines of research and thus rely on methodological pluralism. For example, children with ADHD have academic difficulties, as they have both lower average school marks and score lower on attention span tasks.

Once you have established that there are differences between disciplinary insights, the challenge is to investigate the nature of these differences (after Repko, 2008). You can ask yourself the following questions:

- Do the disciplinary perspectives use the same concept yet mean something different?
- How are concepts defined and measured?
- Do the different disciplinary theories rest on different assumptions?
- Do these assumptions conflict with each other, or can they be complementary?

In Box 1, you see an analysis of the data management tables in Tables 3 and 4. There is tension between the two approaches that focus on the nature or nurture debate as it pertains to aggression. Keeping in mind the above questions, we can see that both papers have a different definition of aggression and a different idea of what amount of alcohol intake is considered problematic. Furthermore, questions arise about the extent to which the findings in animal research can be translated to humans. These starkly contrasting theoretical frameworks seem to have no factor in common at first sight. An interdisciplinary scientist, however, would try to find common ground between these two methodologies by using integration techniques.

Finding or creating common ground by using integration techniques

Once you have gathered insights from all relevant disciplines into the topic you are researching, you can look for common ground between various insights from diverse disciplines to create an interdisciplinary understanding.

Finding common ground is a good basis for interdisciplinary integration and allows you to redefine your main research question in an interdisciplinary way. Remember that such interdisciplinary integration not only involves relevant techniques but can also require creative imagination for the development of a novel explanatory mechanism, intervention, technology, and so on. For more general integration techniques on conceptual and theoretical integration, see Chapter 5.

As explained in Chapter 6, finding common ground can occur along different lines but often involves:
1 Pinpointing a key theory or insight that is shared between disciplines but may be defined and operationalized in diverse ways.
2 Elaborating an explanatory mechanism by integrating additional insights into it.
3 Making assumptions explicit that might need to be reconsidered by one or more disciplines.
4 Realizing that an existing methodology can be improved using insights from other disciplines.
5 Realizing that the apparently contrasting results from different studies can be reinterpreted in such a way that they are consistent with each other.
6 In some cases, common ground is created by an already existing intervention that must be made more robust by adjusting it in response to a newly uncovered additional factor.

Here are some examples of finding common ground. Some sciences share common ground right from the beginning in the form of a comprehensive theory, which might need further elaboration to explain a particular phenomenon. For example, the theory of quantum mechanics is shared between fields as diverse as astronomy, physics, and biology. This common theory can to some extent facilitate their collaboration, as biologists try to explain photosynthetic energy transport using quantum dynamics (Cao et al., 2020) and astronomers use quantum-enhanced detectors to increase the detection rate of gravitational waves (Tse et al., 2019).

Another example is the use of a mechanistic explanation. This allows for the integration of varied types of explanation by extending the meaning of an idea beyond the domain of one discipline into the domain of another. Robert Frank (as reported in Newell, 2007), for instance, extended the meaning of 'self-interest' in economics from its short-term context. He included the long term because, as he argued, someone who acts out of self-interest in the short term may create a reputation that is not beneficial to self-interest in the long term. In other words, he extended the economic meaning of self-interest with insights from sociology and evolutionary biology, enabling him to include certain forms of sacrifice or altruism. Contrasting results such as these can sometimes be made consistent with each other once researchers realize that their focus was on different developmental stages of a phenomenon or that a particular difference between the study populations was in fact influencing the outcomes in an unexpected manner.

Finally, methodologies developed within a certain field may turn out to be valuable in others. For example, remote sensing has yielded many benefits outside geography in fields like archeology, astronomy, ecology, and sociology.

Even if you do find common ground between disciplines, you should not expect the tensions between elements of the different disciplines to now be completely resolved. Instead, it may be useful to focus on these remaining interdisciplinary tensions, as this can lead you to formulate a new question and obtain a novel, more comprehensive insight. Thus, although dissecting the nature of differences may feel like a counterintuitive and even counterproductive approach (would it not be better to focus on where insights overlap or support each other?), focusing on the tensions between research results can provide valuable insights into where common ground can be created.

Especially among the humanities and the social sciences, it is more likely that you must focus on the tensions to find common ground. This has to do with the theoretical and methodological pluralism that reign in these domains more than in others. Harmonious insights are few, whereas conflicting insights are common. Moreover, in addition to their dependence on different epistemological assumptions (i.e., If something exists, how can you know that?), these conflicting insights usually also stem from different ontological assumptions (What can be said to really exist?) (Newell, 2013).

Within sociology, for example, authors like Herbert Marcus and Erich Fromm developed accounts of human society and psychology in which they incorporated both competing Marxist and psychoanalytic theories. They argued that a capitalist society with a strong emphasis on consumerism has an impact on the development of certain psychological attitudes in individuals. To do this, the authors dismantled the theories from their personal claims before they could integrate them into accounts of human society and psychology.

If you are creating common ground by adjusting the assumptions that underlie certain theories or methods, it should be obvious that you may need to reconsider other elements as well. This is because many of the ingredients of a particular science depend on each other – which is captured by the term 'paradigm'. The challenge is therefore to modify concepts or assumptions as little as possible when highlighting hidden commonalities (Newell, 2007).

In sum, insights derived from different disciplines often appear to be incommensurate, conflicting, or diametrically opposed. The foundation of interdisciplinary integration is the awareness that a discipline operates on specific assumptions, concepts, and theories and that these differ from discipline to discipline. The best way to face the challenge of interdisciplinary integration is by being willing to embrace 'the new' and by identifying shared concerns (see examples in Box 2, and practice using Exercise 8.1).

> Box 2
> # Examples of integration in research
>
> **Integrating disciplinary definitions: How to speak 'economish'**
> When Eldar Shafir (Princeton) and Sendhil Mullainathan (Harvard), researchers in psychology and economics respectively, started to work on a theory of poverty, they spent a lot of time creating a common language. Their theory, published in the book *Scarcity – Why Having Too Little Means So Much*, was praised both in and outside academia. On their common language Shafir notes: 'Tolerance, openness, and non-defensiveness of your field are crucial. The goal is to speak the same language, understand each other's issues and perspectives. Take Jerry Fodor and Noam Chomsky. They've been talking to each other for so many years that it's hard to say who's the linguist and who's the philosopher.' (E. Shafir, pers. Comm., 12 December 2013).
>
> As a psychologist, Shafir has invested a lot of time in understanding the economist's perspective. 'I wasn't so much interested in the equations; I was trying to figure out what assumptions they had about people. What is the human agent in the eye of the economist? You have to appreciate the other's perspective to succeed in interdisciplinary work.'
>
> **Connecting theories through a mechanistic model: Sustainable fisheries**
> Managing fisheries was traditionally a matter of catching as many fish as possible. The focus was on controlling the amount of fish in order to secure – or even increase – profit margins. This mode of management can be classified as management-as-control, and its resources (i.e., fish) were considered commodities. Moreover, the ecosystem and the social system were viewed as being completely separate from the fisheries (Berkes, 2003).
>
> This approach inevitably led to the overexploitation of fish stocks, causing collapses in fish populations all over the world. Different stakeholders started to clash when conflicts between ecological, economic, social, and cultural interests emerged (Charles, 1994). Thus, the need arose for a renewed and integrated approach to fishery management. There were no longer two separate systems (fish ecosystem and human social system) but rather one integrated socio-ecological system in which key elements were redefined.
>
> Fish stocks were redefined as ecosystem components with their own niches and functions instead of simply as a commodity. In effect, a concept (which is a component of a theory) was adjusted. Lastly, the management-as-control approach was replaced by an approach focusing on managing for resilience (Berkes, 2003), thus constituting an adjustment of a method. As a result, fisheries have become more sustainable, and the social benefits linked to them have become more stable.

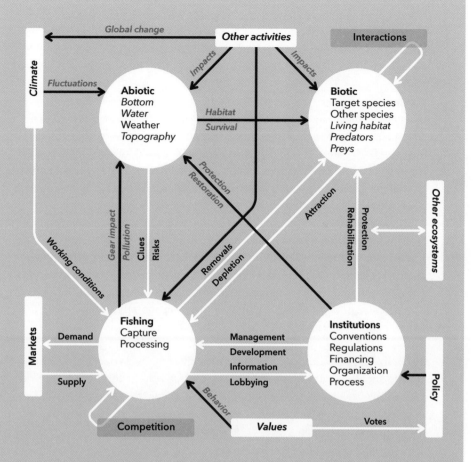

Thus, many problems that require an interdisciplinary research approach encompass a range of interrelated systems/subsystems, mechanisms, and/ or processes. Interdisciplinary research on these problems involves analyzing how these different components are related to each other. A conceptual model can therefore be a valuable tool. In Figure 8, a conceptual model of the socio-ecological system of fisheries is displayed. As you can see, the various subsystems are the different research areas of disciplines ranging from ecology, climatology, and earth sciences to economics, political science, and lawmaking. Their theories are connected, and this visualization makes clear how they are related and how they influence each other.

Combining concepts: detection and reaction of ritualized behaviors

Ritualized behavior – i.e., behavior that is typified by rigidity and repetition – can be found in many facets of life. We see it in cultural rituals but also in the pathological routines of obsessive-compulsive disorder (OCD). The behavior oftentimes lacks a rational motivation, and as a result, a satisfying scientific explanation was lacking.

Pascal Boyer and Pierre Liénard, professors at Washington University in Saint Louis and University of Nevada Las Vegas respectively, combined previous results from evolutionary anthropology, neuropsychology, and neuroimaging to explain ritualized behavior as a 'precaution system' of the self (Boyer & Liénard, 2006).

From an anthropological perspective, they distilled the similarities between cultural ritualized behavior and individual pathology. The person performing the ritual feels compelled to do it and does so in a rigid format. Often, both types of rituals involve much repetition and redundancy, thereby demoting the goal of the ritual. Many rituals are also centered around certain themes such as purification and protection.

From neuropsychology and neuroimaging, Boyer and Liénard knew that when a person's anxiety levels are high, the brain tends to focus on low-level gestural units rather than on the higher-level units that would relate to the goal. Thus, the working memory gets overwhelmed by low-level units. This brings a temporary comfort to the performer of the ritual, thereby strengthening the apparent idea of its necessity.

By combining these concepts of anthropology and neuroscience, Boyer and Liénard showed that activation of this 'precaution system' through anxiety, for example, can lead to ritualized behavior and thus the pathology of OCD. Activation of this system in a larger society also explains why cultural ritualized behavior is so compelling to large groups of people.

Developing interdisciplinary technologies: a social robot for children with autism

To help children with autism, computer scientists Kersting Dautenhahn, Ben Robins, and their team at the University of Hertfordshire developed a humanoid robot, Kaspar (Dautenhahn et al., 2009). Kaspar can be used in therapy sessions as a learning and teaching tool to facilitate social interaction skills. This project is a great example of a transdisciplinary research, where computer scientists worked together with people from other academic disciplines but also received valuable input from teachers, health professionals, and parents to make the robot more effective.

Before they knew what type of robot to design, they had to know how to specifically cater to the needs of autistic children. Their research proved there is an immense heterogeneity in such needs, which meant that they should aim to design a very versatile robot or a range of robots. Multiple types of robots were tested to find the best shape to serve as a therapeutic toy before Kaspar was selected. Further research on the interaction between children and the robot served as input for improvements and underlined the iterative character of research in general and transdisciplinary research in specific.

Step 5   Finalize interdisciplinary research question and sub-questions

Using one or more integration techniques, you have so far identified a useful overlap between disciplines, developed an integrated theoretical framework, and identified gaps of knowledge on a specific topic. Based on these insights, you will be able to formulate an interdisciplinary research question, which will be the core of your research project.

Along the way, you will find new valuable articles that help to specify and refine your data management table, which may prompt you to reformulate your research question. As mentioned before, interdisciplinary research is an iterative process, so it is likely that you will return to your research question and adjust it after each step.

To finalize the research question, make sure it is relevant, precise, researchable, and anchored (for more details on these criteria, see Chapter 7). Once you have finalized the main research question, you will have to divide this into sub-questions. Although the same criteria apply here as for the main research question, there are additional factors to consider:

- It is important that the intended answers to the sub-questions *together* lead to an answer to the main research question, so adjust your sub-questions until they cover the finalized research question completely.
- Make sure that your sub-questions are logical steps that lead to an answer to the main research question. For example, you can organize the answers to each sub-question in a separate paragraph.
- Although interdisciplinary projects often contain sub-questions that give a monodisciplinary perspective on your main research question, the integration of disciplines can also take place *within* a sub-question.

When dividing your research question into sub-questions, you can decide to divide a concept into terms that can be measured more easily. In the example on fogponics (Chapter 11), the finalized research question 'To what extent can fogponics contribute to a more sustainable form of greenhouse cultivation?' is subdivided into two questions:

- Is fogponics a better alternative in terms of efficient nutrient use when compared to conventional greenhouse agriculture?
- Is fogponics a better alternative in terms of efficient water use when compared to conventional greenhouse agriculture?

The sub-questions themselves can be divided into questions. In the example of fogponics, one could research per nutrient whether fogponics or conventional greenhouse agriculture is a better alternative. However, be careful not to go overboard in subdividing your sub-questions, as this can lead to results that are redundant or too specific.

Exercises

8.1 Reflection on assumptions within theoretical framework

It is important to reflect on each other's disciplinary assumptions, as this will not only enhance the communication in your interdisciplinary team but also resolve (seemingly) contrasting theoretical outcomes that may arise in your theoretical framework.

For this exercise, each team member first answers the following questions privately. After answering these questions, the outcome should be discussed within your team.

1. Give a clear definition of the main concept of your research project (e.g., the concept of 'evolution' can be defined as 'genetic change in a population').
2. Describe how to test assumptions about this concept (e.g., How is evolution verified or falsified in a certain system?).
3. Then ask whether this type of evidence can be questioned (e.g., Is a genetic change the only way an evolving system can change?).

For further discussions on disciplinary assumptions – e.g., on levels other than the conceptual level (such as the research method or ideas about data analysis) – choose three questions from this toolbox (scan the QR code) and discuss the individual outcomes within your interdisciplinary team.

9 Data collection and analysis

After reviewing the relevant literature from the disciplines that were identified as being essential for addressing your initial research question, you have developed a theoretical framework that enables you to refine the research question. Now the question is: how are you going to answer this refined research question and sub-questions? It is time to develop a methodological framework for data collection and data analysis.

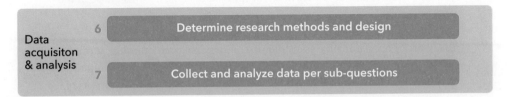

Figure 18   Steps 6 and 7 of the interdisciplinary research model. A methodology is designed to collect and analyze data.

Considerations:
- What are the relevant methods each discipline has to offer?
- Is a synthesis of disciplinary methods possible?
- How does the choice of methods influence the anticipated results?

The process of developing such a methodological framework to structure the practical matters of your research project is similar to the process you used to develop a theoretical framework to define the theoretical context and to specify the focus of your research. This methodological framework is what we call 'the design' of your research. The process of translating your research questions into measurable, researchable dimensions is what we call 'operationalization', which was already discussed in Chapter 2 in the context of the science cycle.

In this chapter, we will guide you through the process of developing an adequate and manageable interdisciplinary methodological framework. To do so, we will take you through the following guiding questions:
- What (kind of) information is required to answer the research question?
- What approach is the most appropriate for answering the sub-questions?
- Which methods are the most appropriate for producing the data needed?

Step 6   Determine research methods and design
What (kind of) information is required to answer the research question?

The sub-questions you are trying to answer can take many forms. Therefore, it is useful to keep in mind what kind of question you are asking: is it an observational or an experimental question? Or does it lead to practical problem-solving? The starting point can also be a prediction or a hypothesis instead of a question. Each of these questions calls for a different anticipated result. Keep this anticipated result in mind when designing a methodological framework.

In the theoretical framework, you explored the concepts and relations related to your research questions. For investigating the questions, you now need to comprehend what you want to be measure: you need to operationalize them in order to manipulate, test, or explore them further. In other words, you are developing one or more ways of translating relevant concepts into researchable and measurable items (see also Chapter 2). Since interdisciplinary research often involves exploring new fields, existing methods are not always sufficient to operationalize your research question. That is why creativity is very important also during this phase of the research project.

To creatively operationalize your interdisciplinary research question, it can be very useful, if not essential, to make an operationalization scheme per sub-question (see also Exercise 9.1). In this scheme, you should set out: (1) the concepts of your theoretical framework, (2) the measurable dimensions of the concept, (3) indicators on how to measure the dimension, and (4) the variables, or the units in which you will measure these indicators. For example, if you wish to research the relationship between deforestation and quality of life for indigenous people in Brazil, the concepts described in your theoretical framework would include biodiversity, standard of living, and the retaining of their indigenous culture. Biodiversity could be measured in species richness or species evenness (dimensions), while for standard of living you could look at the gross income per capita. To measure the involvement with indigenous culture, you could look at the cultural attachment of people to language and cultural events, for example (Table 5).

Although operationalization, dimensions, indicators, and variables are not the kinds of terms that you might typically associate with research in the humanities, such research does involve similar elements. For example, for the interpretation of a particular religious sacrifice, choices must be made that together imply a form of operationalization. These choices help to make the research manageable, since they are effectively constraining the research task. What are the relevant dimensions or features of the sacrifice that need further exploration, what other features might be – at least for the moment – be considered less relevant? What resources might be consulted in order to find reliable and relevant 'indicators' for this interpretation? Whether humanities scholars fill in such an operationalization scheme or not, they take steps that are similar to these.

Table 5 Example of an operationalization scheme

Concept	Dimensions	Indicators	Variables
Biodiversity	Species richness	The amount of species living in a certain region	Species/km^2
	Species evenness	The evenness between number of individuals between species	Shannon Index (Shannon, 1948)
Standard of living	Gross income per capita	The amount of income per person	R$/person
	Life satisfaction	Self-reported judgment of one's life satisfaction	Satisfaction with Life scale (Diener et al., 1985)
Indigenous culture	Cultural attachment (after Dockery, 2013)	Speaks indigenous language	% of people
		Self-reported importance of attending cultural events	Three point Likert scale with 1 = not important, 3 = very important
		Self-reported identification with culture	Five point Likert scale with 1 = strongly disagree, 5 = strongly agree

After completing the operationalization scheme, it is important to check if all your indicators measure what you want to measure and all aspects of your sub-questions are covered. With this scheme, you now have a list of variables you need to collect in order to answer your sub-questions. Your next step will be to determine how to collect the required information.

What approach is the most appropriate for answering the sub-questions?

How to devise a useful methodology for collecting the information you need depends on your approach. As we saw in Chapter 1, there are different ways of thinking about knowledge and how knowledge should and can be gained. The approach you take can be positivistic and quantitative or interpretative and qualitative, for example. Especially for interdisciplinary and transdisciplinary research, you may want to opt for an integrated combination of both (i.e., a mixed-methods approach, or a cooperative method approach). In many cases, it is only through a combination of approaches that we are able to get to a more complete understanding of our topic, since we are often investigating characteristics (of a phenomenon) that interact with each other, making them difficult to determine when using non-integrated methods (see also methodological pluralism in Chapter 2). And once we have applied such methods, we must face the challenge of dealing with a new integrated data type as well.

The approach you use also depends on the kind of question you ask. In the case of interdisciplinary research, it is often worthwhile considering using different approaches at the same time to cut across disciplines and underpin a holistic conclusion. This allows you to obtain a more robust insight or explanation, as it is confirmed by different approaches using different research methods (see Chapter 5).

Which methods are the most appropriate for producing the data needed?

There are different ways to find an answer to your research question, and the answer you find will depend, to some extent, on the operationalization you chose and the methods and techniques you used. For example, the results of a specific qualitative case study will differ in nature from a quantitative comparative study into the same topic, but the key point is that there will be some kind of a relation between the two sets of results. An interdisciplinary goal would be to integrate these results irrespective of their methodological differences: the qualitative insights might help us to adjust our questionnaires and thus contribute to enhanced quantitative research.

To find out what the customary methods and techniques are in different disciplines, it is wise to return to your data management table. The table provides a brief overview of the articles you have read and analyzed so far, and from these articles you can learn how the key concepts were operationalized in the different methodological frameworks. When looking at the methodology of disciplinary studies, you should decide how to adapt or combine the methodologies used when conducting your own research.

Within disciplines, preferences for specific methods are common. But if you conduct research into a problem that crosses disciplinary boundaries, a combination of techniques often offers more accurate results. This can have the effect of lifting your research from the level of multidisciplinarity to that of interdisciplinarity or transdisciplinarity. There are multiple potential configurations to cross disciplinary boundaries in methodology:

1 Expanding one discipline's method to cover another discipline's sub-question, e.g., the adjustment of a specific theory to use it as input for a data gathering technique in another field.
2 Crossing over of methods. You might consider using your knowledge of and skills in a disciplinary method towards a particular problem that is new to the method.

3   Combining methods. You can use the results of method A in method B or use multiple methods on the same problem to meet the requirements set by the research problem and rule out another method's limitations, thus yielding a more robust result (see Chapter 2).

Box 3 provides examples of student projects that used integration in the methodological framework.

The choice of method, the practical limitations of methods, and the time allotted for your research may force you to reconsider your research question and adapt it to fit these limitations. This is part of the iterative process that is inherent to interdisciplinary research. Adapting to the limitations is challenging but unavoidable, so do not get discouraged too easily. Just be aware that you may have to modify your research question as a result of practical limitations and keep in mind that this generally means you are moving forwards, not backwards!

> **Box 3**
> ## Examples of integration at the level of methods
>
> **Measuring the weight of the dodo**
> In historical images of the dodo, this extinct bird is sometimes pictured as being rather heavy and at other times as being lightweight. In reality, its weight was unknown. In compiling their data management table, four students (van Dierendonck, van Egmond, ten Hagen, & Kreuning, 2013) found that many different methods have been used to reconstruct the weight of the dodo. They studied research that had been conducted in the past and found differences in the reference species that was used (in other words: the non-extinct bird species the dodo was compared to), differences in the type of bones that were used as an indicator to determine the overall weight of the animal, and differences in which part of the bone was measured. The students integrated elements of these research methods and their underlying assumptions into a new method, with which they were able to assess the dodo's weight more accurately.
>
> **Volunteer tourism in Cusco**
> During a research project on volunteer tourism in Cusco, an interdisciplinary social scientist (Schram, 2012) wanted to learn more about how the different actors in volunteer tourism experienced power relations as well as how they evaluate the dependencies they experience. Interviewing relevant actors seemed an appropriate research method.
>
> However, thoughts and feelings about power relations are often part of a larger narrative. Moreover, individuals are not always aware of their attitude toward it. It is therefore difficult – if not impossible – to get relevant answers via actor interviews about this topic.

Positioning theory (from the field of social psychology) gives an explanation for how people define the 'self' in conversations. In his research project, Schram used the knowledge gained from positioning theory to improve the interview method. Respondents were asked to write down on post-it notes the different groups of actors they believed were involved in volunteer tourism in Cusco. They were then asked to categorize the notes in two different ways: from least to most influential, and from dependent to independent. In this way, positioning theory was used as a method in the domain of social interactions, and it was thereby possible to gain insights into the way individuals perceived power relations.

Omega-3 fatty acid and heart rate variability
In order to find the impact of omega-3 fatty acid intake on heart rate variability (HRV) in men and women, two students with backgrounds in biomedical sciences and mathematics realized that they first had to look at how HRV was being analyzed. They found the current technique to be outdated and therefore introduced a new one: the mathematical logarithm ApEn. This new technique and the use of an existing dataset allowed them to determine the influence of omega-3 fatty acid intake on HRV in men and women with more accuracy. In this project, the analysis method of one discipline (mathematics) affected the data acquisition of another discipline (cardiology), as a slightly different dataset was needed for the mathematicians than would normally be produced by monodisciplinary cardiology research (Bekius & Elsenburg, 2010).

Step 7   Collect and analyze data

Now that you have crafted an integrated methodological framework, you are ready to collect and analyze your data. When you select a method, you are also committing yourself to the accepted criteria and standards for collecting and analyzing data with that method. If instead you have translated or adjusted a method or technique, this may mean that you will need to reconsider your methods of data collection and analysis and possibly adjust them – but in such a way that does not negatively affect the validity, reliability, and accuracy of the method (see the Omega-3 fatty acid and heart rate variability example in Box 3).

Data collection

The first step in implementing the methodological framework is to collect the necessary data. It is important to check the accuracy of the collected data along the way to ensure that your end results are trustworthy. There are two forms of data collection: primary data collection is where you generate the necessary data yourself, and secondary data collection is where you use data generated by others (Figure 19).

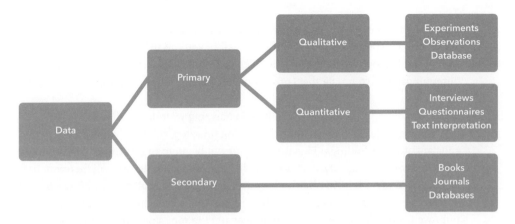

Figure 19   Data types with examples

Primary data collection can be split up into qualitative and quantitative research methods. The method chosen depends on your approach and research question, as discussed in Step 6. Quantitative numerical data can, for example, be gathered by conducting experiments, making observations, or by curating your own database. Qualitative data, which is any data that is non-numerical, can be gathered through interviews, text interpretations, questionnaires, surveys, or observations, to name a few examples. Whichever method of data collection is used, you must make sure the integrity of the research question is maintained. For example, experimental data should be reproducible, and data should be checked for validity.

In secondary data collection, checking your sources for validity and reliability is of the utmost importance. When you use data from other researchers or organizations – whether these are books, journals, documents, or an existing database – you must always check the background of the authors, the publication year, the contribution of the piece to the existing science, and any conflicts of interests.

Data analysis

With all the necessary data collected, it is now time to analyze it. A first step in data analysis is checking the raw data. Some of your data may simply be false or not useful (for example as a result of a mistake in an experiment). Suppose that you used survey questionnaires as a method for data collection and asked participants to fill in their age. If somebody put down '763 years old', you know this cannot be true and you must adjust the otherwise skewed results.

In some cases, you need to categorize or code your data. There are different techniques for coding data and, as mentioned previously, the technique you choose depends on the kind of data you collected and the goal and approach of your research. If you use a quantitative approach, your data should be coded in a way that allows you to perform statistical analysis. For example, if you asked participants about work satisfaction and the possible answers to choose from were 'not satisfied', 'more or less satisfied', and 'very satisfied', you need to label these categories with different levels before you can use them in statistical analyses. If you use a qualitative approach, you should also code your data. Before you can interpret the dataset, you should think about the themes or variables you want to use to analyze and mark the different parts of your data that correspond with that theme or variable.

Upon organizing your data, you can begin analyzing your data. A first step is to plot out your data for a visual analysis. During this step, statistical analysis is key to substantiate your findings. This is when you will begin to find correlations in your data. The manner of analysis very much depends on your data and will therefore not be discussed here. It is possible and even advisable to use multiple techniques to analyze the same data. Integrating different data analysis techniques might actually improve the quality of your research project and the robustness of its outcomes.

Exercises

9.1 Operationalization scheme tool

There are three steps to making an operationalization scheme:

1 Identify the concepts and dimensions

Examine your research question(s) and theoretical framework. To start your operationalization scheme, list all concepts stated there. In some cases, you will need to further specify these concepts; this will depend on your definition of the concept and may also depend on your discipline. A concept can exist in multiple dimensions, and hence it is important to collaborate with your research team to come up with an appropriate definition of the concept by defining its dimensions.

2 Select indicators that represent the concepts

To measure the concepts, you must translate them into measurable indicators. For some concepts, these indicators will be obvious or standardized, but for other concepts, this will be more difficult to measure. This can be a particular challenge with concepts that span multiple disciplines. You can either look at the concepts from multiple angles by using more than one indicator or try to combine them into one overarching indicator.

3 Choose the variables to measure the indicators

Now that you have chosen what you will measure, it is time to choose how you will measure them, and in which units or scale. This is a crucial step, as this will determine what your results will look like. The best way to do this is to think about the preferred form of the expected results and decide which variables will get you there. This determines whether your approach will be qualitative or quantitative.

10 Completion of your research

It is common for readers of research articles to jump immediately to the final sections – the discussion and the conclusion – after having read the introduction, eager to know the results and insights from the study. These final sections are typically not meant to present the reader with new information but rather to help them draw conclusions and discover the study's limitations and questions for future research.

Figure 20   Steps 8 and 9 of the interdisciplinary research model. The analyzed data is first integrated, interpreted, and discussed before a conclusion is drawn.

Considerations:
- Do the results confirm your hypothesis or expectations from your integrated theoretical framework?
- Does your interdisciplinary insight shed new light on the insights obtained by each discipline separately?
- Be explicit about how specific disciplinary insights have been integrated into your interdisciplinary conclusion and how this might have an impact on future disciplinary research.
- What parts of your interdisciplinary research may be sensitive to criticism, and why?
- Based on your results, what knowledge gaps remain, and what research question would you want to pose next?

Step 8 Integrate results related to the sub-questions

In order to interpret your results in a way that will answer your research question, it is crucial that you integrate the results related to your sub-questions (see Chapter 5). If you used multiple disciplinary methods to answer an interdisciplinary question, you will have to integrate the different datasets by looking at the data obtained via one method in the light of the data obtained from another. Similarly, when you are looking at an optimization problem such as current climate models, it is usually necessary to integrate the different subsets of data into a more comprehensive model. Furthermore, the integration of results always occurs naturally when the research leads to practical solutions or products (like the social robot in Box 2).

To convey your integrated results effectively to your reader, you should think carefully about the way you will present or visualize the data or findings. This can be done by means of graphs, tables, infographics, timelines, or any number of ways. Note that each discipline has its own norms when it comes to visualizing data. Especially when carrying out interdisciplinary research as a team, it is crucial that you effectively communicate the integrated results not only to the outside world but also within the research team. It is thus important to look for the optimal way of visualizing your data to make it understandable across the involved disciplines. Moreover, this information should help you to answer your sub-questions clearly in your discussion and conclusion.

A good way to visualize and integrate your results at the same time is by creating a conceptual model (such as the example on fisheries in Box 2), which allows you to present the system or problem you have researched in an orderly fashion. Subsequently, you can uncover interactions between different aspects of the problem that are typically addressed by different disciplines and thereby expose leverage points that you can address in the discussion.

Step 9 Interpret results, discuss research, draw conclusions

In your discussion, you interpret your results by putting them into the context of the existing literature. You explain how your answers fill a knowledge or reasoning gap and how they concur or contradict the state of the art. You also discuss possible alternative interpretations of your results as well as the limitations of your research. Then, in the conclusion, you lay out the arguments that you have built up over the research. The conclusions of your research project are likely to raise new questions. Based on those new questions, you can give your recommendations for future research.

Interpret results

The most important task of your discussion section is to interpret and discuss your results. Keep in mind that different disciplines have different standards for how to write a discussion. A philosophical paper that tries to bridge a gap in reasoning may wholly consist of a discussion of different viewpoints, while a natural sciences paper trying to fill a gap in knowledge will use the discussion section to discuss results in a greater context and will include the implications and limitations of the paper. Be mindful of the different gaps you are tackling, and make sure you communicate with your colleagues on how to structure your discussion section.

A typical discussion will start with a short overview of the results and how they answer the main research question. Next, this answer is evaluated in relation to the theoretical framework and relevant literature. Even if the results completely confirm the hypothesis, you can always discuss your choices and interpretations. As you may recall from Chapter 2, logical induction is always limited, and observations are inevitably theory-laden. As a result, confirmations should always be treated with caution. Your results are often supported by previous research, but it is likely that there have been studies that do not necessarily agree with your conclusions. Do not ignore those alternative explanations; instead, give them a place in your discussion, as it might offer options for future research by other researchers.

Discuss research

Besides interpreting your results, the discussion section is the place to reflect on the choices you made during your research. At the beginning of the research project, you started with a problem, which you then narrowed down to a research question and sub-questions. Subsequently, you operationalized a way to investigate the question for which you then chose appropriate methods and instruments. In this process of specifying the problem, you inevitably made several choices, which may have forced you to reconsider some of the choices you made in previous steps. In order to formulate a well-considered discussion of your research, you need to reflect on all the choices you made throughout the project that might have influenced the results. Articulate the limitations of your interdisciplinary research project and reflect upon the limitations of your study's methods. Suggest possible alternatives and follow-up studies as you draw your discussion to a close.

During this reflection, you must take great care not to lose sight of the interdisciplinary character of your research (see Exercise 10.1). The need and implications of this method must be addressed. Ask yourself why this particular research question required an interdisciplinary perspective – i.e., what would the shortcomings of a disciplinary perspective likely have been? Also take a moment to consider the implications that your research and conclusion have for those disciplinary perspectives involved in your study. What advice would you give other disciplinary or interdisciplinary researchers who want to contribute to this field? After reflecting on your own research, reflect on what your research does *not* answer: what gaps in knowledge or reasoning remain, and what gaps have emerged as a result of the insights provided by your research?

Draw conclusions

At this point, you have taken all the steps that eventually led to an interdisciplinary answer to your integrated research question. You have brought together different disciplinary insights and integrated them to get a more complete and inclusive understanding of the complex problem you have researched. It is now time to conclude your research.

In the conclusion, you should systematically reflect on your research and lay out the arguments you have built up over the rest of your research. While reflecting on your arguments, place your conclusions in the context of the bigger picture and future perspectives. Note that this is more complex in the case of interdisciplinary science, as you not only address future interdisciplinary research projects but also inspire disciplinary sciences that relate to your research topic. Therefore, it is important to remind yourself that you are not targeting a specific audience but rather appealing to a diverse audience consisting of both disciplinary and interdisciplinary academics as well as professionals.

In transdisciplinary research (see Chapter 5), the end product is not always a scientific report but may also be a product (e.g., a new technology), policy advice, or a tool (e.g., a new form of therapy). In this case, the requirements for a conclusion are more practical in nature and may be a call for action. Keep in mind that transdisciplinary conclusions have many implications for society, like political choices and social values. Therefore, you should be extra careful and explicit when formulating conclusions with a large possible impact.

The conclusions of your research project are likely to raise new questions. Based on those new questions, you can give recommendations for future interdisciplinary or transdisciplinary research. There are multiple directions you can go with this. For example, you can suggest that future researchers improve on the current results with more disciplinary research or use the findings of your research in a new interdisciplinary project.

The last sentence of your paper is one that readers will remember, as it is the last thing they read. It is therefore an important one and as such needs careful crafting. A good option is to conclude with your take home message: a clear and concise answer to your main research question that stresses the importance of your project.

Exercises

10.1 Interdisciplinary reflection

To reflect on the interdisciplinary research process and the surplus value of your interdisciplinary approach, you can ask yourself the following questions:

- Why did you need an interdisciplinary perspective – i.e., what would the shortcomings of disciplinary perspectives likely have been?
- Did the interdisciplinary research process lead to unexpected insights? If so, where did they occur?
- Were these unexpected insights related to the main research question or tangential to your line of inquiry?
- Were there steps specified in the interdisciplinary research model we provided that did not fit the project well? If so, how did you adjust the process?
- Overall, what did you learn about the interdisciplinary process?
- Do your interdisciplinary insights shed new light on the individual insights from each discipline? Or can your insights be criticized for lacking depth or being too reductionist?
- Has your analysis perhaps unearthed some important issues that deserve more attention?
- What gaps in knowledge remain and, more importantly, what gaps in knowledge have emerged as a result of the insights provided by your research?

Part 3
Interdisciplinary research in practice

11 Interdisciplinary research example

In this chapter, we take a close look at an interdisciplinary capstone research project carried out by a group of five bachelor's students at the University of Amsterdam, the Netherlands on fogponics, a promising and innovative way to grow crops (Bakker, van der Linden, Steenbrink, Stuut & Veldhuyzen van Zanten, 2014). Using our interdisciplinary research model, we describe their interdisciplinary research process step by step.

Step 1   Determine the problem or topic

The five students were interested in sustainable agriculture, and after some brainstorming and an initial literature search, they came up with an innovative way of growing crops sustainably. On the scale at which it is currently being practiced, agriculture is an unsustainable form of land use because the excessive energy use, water consumption and pollution (from fertilizers, pesticides, herbicides, etc.) damage the environment and humans. Moreover, the global human population is likely to continue to grow in the coming decades, which will further increase the demand for food. To cope with these challenges, new, efficient, and sustainable forms of agriculture must be developed. Vertical farming (agriculture on vertically inclined surfaces such as a skyscraper) might contribute to solving the problems related to current greenhouse agriculture, which is especially large in the Netherlands. Fogponics, a method of growing crops using a nutrient-rich fog, might be a suitable and sustainable way to grow crops in a vertical farming setting. This research will focus on the situation in the Netherlands, where much information on this innovation is available.

Relevant disciplines

For this project, the relevant disciplines were partly determined by the disciplines that the group of students were studying. Other disciplines could have been relevant with regards to this topic, but the final research question was defined in such a way that the following contributing disciplines would be able to answer it:
- Biology (plants)
- Chemistry (nutrients)
- Artificial Intelligence (system operating)
- Mathematics (extrapolation of calculations)
- Economics (costs)

Dominant perspectives (with regard to the problem and the contemplated solution)
- Biology: Agriculture is unsustainable, as pollution is damaging humans and their environment. Fogponics might contribute to a more sustainable form of agriculture. Biology can shed light on the physiological aspects of the growing of crops within a fogponics context.
- Chemistry: Agriculture is unsustainable, as it requires significant energy inputs while fossil fuels are becoming increasingly scarce. For fogponics to be a sustainable alternative, knowledge of chemistry is needed to make the use of nutrients and water as efficient as possible.
- Artificial Intelligence (AI): AI focuses on 'intelligent' systems. In this case, AI can help to design a fogponic system that operates autonomously. Furthermore, it can help to optimize this system when implemented on a larger scale.
- Mathematics: Mathematics has no perspective on the topic, but will be a vital part of the technical exploration of the solution to the problem.
- Economics: By 2050, the world population will have outgrown the maximum production of the current agriculture system. Hence, from an economic perspective, our current system will not be sustainable or cost-efficient in the long term. Economics can also help to assess the cost-efficiency of fogponics on a larger scale.

Step 2 Formulate preliminary interdisciplinary research question

By integrating the relevant perspectives of their respective disciplines, the team of students formulated the following research question: To what extent can fogponics contribute to sustainable agriculture?

Relevant theories, concepts, and assumptions

Looking for relevant contributions from their respective disciplines to answer the main research question, the students subsequently formulated potential disciplinary distributions. Biology, they contended, can shed light on the physiological aspects of crops, and it can also help to select the optimal crop species for an experimental fogponic setting. Chemistry is relevant for measuring and analyzing nutrients used by the plants. By using mathematical models, an extrapolation can be made from the experimental level to the national level. And a comparison can be made based on economic insights between fogponic systems and current greenhouses in terms of the use of water and nutrients.

Step 3 Identify relevant theories for each discipline

In rocky habitats close to waterfalls, plants grow with roots that 'hang' in the fog generated by the falling water. This observation inspired plant biologists to develop aeroponic systems (systems where 'hanging' roots are sprinkled with water) to study root functioning. Aeroponic systems were subsequently developed into a setup where water is provided in the form of fog, a so-called fogponic system. This research project was inspired by such a fogponic system, with the idea that it might also be used in a commercial crop production setting. Tomatoes (*Solanum lycopersicum*) are the world's second largest crop in terms of production. Additionally, tomatoes are an important export product for the Netherlands, which is currently the largest tomato-exporting country in the world. These were two arguments for the students to use tomatoes as a model organism in the fogponic system.

Chemistry can help measure the amount of nutrients used by the plant. This can be analyzed by means of high-performance liquid chromatography (HPLC), a method to identify and quantify the components of a mixture. There is a high variance in the number of cations (positively charged ions) and a low variance in the number of anions (negatively charged ions) in the nutrient water. Using a cation-exchange HPLC, one can measure the number of cations in the liquid. These cations represent nutrients, which means the amount of nutrients can be measured.

To compare fogponics with current greenhouse agriculture in terms of nutrient and water use, it is essential to assess the costs, production capacity, and use of the resources (i.e., nutrients and water) of fogponics in the long term. This can be done using the relationship between plant growth and water and nutrient use. Plant growth can be simulated by means of a mathematical growth model.

To make a fogponic system as efficient as possible, artificial intelligence (AI) can provide useful tools. AI can develop an expert system to regulate the circulation of the water as well as the concentrations of nutrients in the water. Such a system would be semi-autonomous and would have the ability to learn, thereby improving efficiency. Sustainability and efficiency must go hand in hand to be consistent with the main research question in this project.

The World Commission on Environment and Development (1987) defined sustainable development as 'development which meets the needs of the present without compromising the ability of future generations to meet their own needs'. In this research project, sustainability was defined as an interdisciplinary concept, encompassing both environmental and economic stakes or needs.

Step 4   Identify overlap and distinctions between different perspectives

Although more topics of overlap can be identified in this project, one clear overlap between all student disciplines was the topic of 'resource efficiency'. Biological systems comparable to fogponics have been known to occur in the natural world (e.g., hanging plants in humid areas), using only water and nutrients from the air moisture. This shows an unexplored ecological niche which agriculture can benefit from, by imitating the resource efficiency of these natural plants in a greenhouse. Furthermore, the resource efficiency is optimized within the system itself. By implementing a self-learning (AI) system using fogponics, tomato growth can be further increased, while a chemical method developed by the project allows nutrient concentrations to be easily quantified.

A clear distinction between the different student disciplines may lie in their levels of organization, the hierarchical complexity of the operationalized concepts. For the chemist, the main focus lies in quantifying the plant's nutrient concentration, but another perspective might prioritize other factors. For example, when comparing the fogponic system with more conventional systems, tomato yield rather than nutrient concentration might be more important.

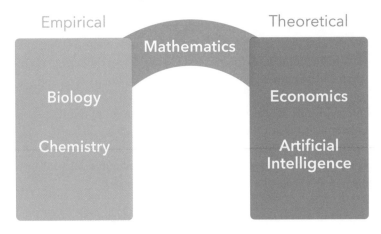

Figure 21 Schematic representation of the relation between the disciplines contributing to the research (Bakker et al., 2014)

Step 5 Finalize interdisciplinary research question and sub-questions

The final research question was somewhat different from the preliminary research question, as the students decided under what specific conditions they would experiment with fogponics. As a result, the final clause was changed and specified, leading to the question: To what extent can fogponics contribute to a more sustainable form of greenhouse cultivation? As explained earlier, this was done by means of a case study on greenhouse production of tomatoes in the Netherlands.

Given the limitations inherent to a student project and the constraints determined by the disciplines involved in the project, the students had to limit and specify their sub-questions too. They decided to focus on nutrient and water use, as they are highly relevant for any agricultural project's sustainability. As a result, the research question was subdivided into two sub-questions:

- Is fogponics a good alternative in terms of efficient nutrient use to conventional greenhouse agriculture?
- Is fogponics a good alternative in terms of efficient water use to conventional greenhouse agriculture?

Step 6 Determine research methods and design

A combination of both empirical and theoretical research was used in this research project, with the different disciplines playing key roles. This is illustrated in Figure 21. The empirical data on plant growth in a fogponic system were collected using an experimental setting, which involved the disciplines of biology (specifically, plant biology) and chemistry. The theoretical research consisted of a study on the economics of tomato greenhouse cultivation and another study on how to optimize the fogponic system (using knowledge from AI). The data from the experiment were used as input for the mathematical extrapolation, on the basis of which the theoretical part of this research was carried out. As you can see, the methods were integrated, making the research truly interdisciplinary.

During the experiment, daily samples of the water were taken, and nutrient concentrations were measured with HPLC. Deficits in nutrient concentrations were replenished. Due to the limited duration of the project, the tomato plants did not reach maturity during the course of the experiment. Therefore, to assess the amount of water and nutrients needed by full-grown plants, the total length of all stems of the tomato plants were measured and used to create a growth curve (Figure 23).

This growth curve formed the basis for an extrapolation of water and nutrient use. The next thing the students calculated was the hypothetical amount of water and nutrients that the entire Dutch tomato greenhouse sector would use based on the fogponic method. Subsequently, this was compared to economic data on conventional greenhouse cultivation (water use, nutrient use, and costs), which was obtained through literature research.

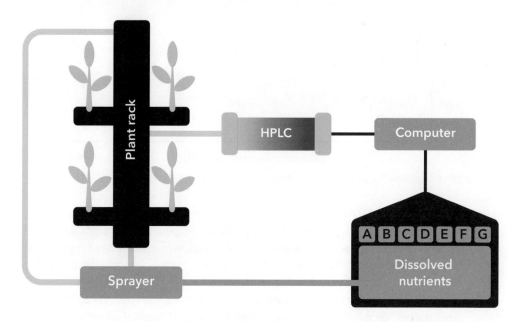

Figure 22   Experimental fogponics system (Bakker et al., 2014)

Step 7 Collect and analyze data per sub-question

To answer the main research question – 'To what extent can fogponics contribute to a more sustainable form of greenhouse cultivation?' – plant measurements had to be extrapolated to potential greenhouse tomato yields at the national level. Plant growth was measured for approximately 25 days. The results of the plant growth measurements were extrapolated using an S-growth curve, because this best describes the growth of tomato plants. An extrapolated graph was created by means of a logistic growth function (see Figure 23), which displays the number of days on the x-axis and the total plant length (including all stems) on the y-axis. Using a logistic growth function, the students were able to determine that the maximum length the tomato plants would reach in the fogponic setting is about 600 cm.

Subsequently, the amount of nutrients and water needed for full-grown plants was calculated. These amounts were compared to those needed in the current greenhouse tomato sector in the Netherlands. The students found that the fogponic system requires approximately 30% of the water and just 14% of the nutrients used in conventional greenhouse cultivation of tomatoes. This, in turn, would lead to a yearly reduction in costs of around €7 million in the Netherlands.

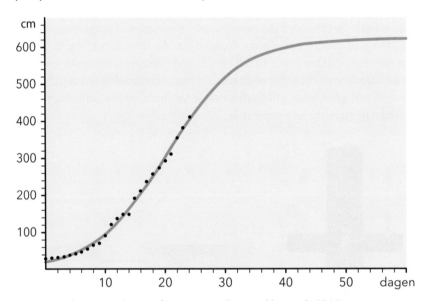

Figure 23   Extrapolated growth curve from tomato plants (Bakker et al., 2014)

Step 8 Integrate results related to the sub-questions

Compared to conventional greenhouse agriculture, fogponics is more efficient in terms of both water use and nutrient use. Consequently, fogponics has the potential to significantly reduce the costs of Dutch tomato production: by approximately 85%, to be specific. Therefore, both sub-questions can be answered affirmatively. The project concluded that fogponics outperforms conventional greenhouse agriculture from the perspectives of the biological, chemical, and economical disciplines, supported by AI technology and mathematics.

Step 9 Interpret results, discuss research, and draw conclusion(s)

Discussion

The results confirm the hypothesis that fogponics can contribute to sustainable agriculture, both in terms of environmental and economic needs. These promising results, together with the interdisciplinary insight into the valuable combination of fogponics with commercial agriculture, create possibilities for new developments in sustainable agriculture. This interdisciplinary research, in turn, raises several new research questions. For example, in this project, sustainability was demarcated to water and nutrient use and costs. However, there are other aspects of sustainability that were not considered. Future research focusing on such aspects as energy consumption, resource use, and waste production can lead to a more complete picture of the sustainability of fogponics.

The students realized that a few aspects might have undesirably influenced the results of the research. First, the tubes in which the tomato plants were placed were not completely closed. Consequently, a leakage of fog occurred. It is unclear how much water and nutrients leaked away from the system. Moreover, the potential negative effects of a completely closed system on plant growth cannot be ruled out. Second, due to the limitations of the research project, it was necessary to extrapolate the data derived from a small, nonreplicated experiment. To make the results more robust, the experiment should be repeated, and the experiment time extended so that the tomato plants can produce their entire yield. Third, several assumptions were made concerning the production of tomato plants. These assumptions were partly based on literature and interviews with experts. Nonetheless, these numbers (growth rate means, etc.) might not correspond precisely with reality, as different, unidentified factors may also influence the results.

Conclusion

The conclusion can be drawn that fogponics is indeed more sustainable – from an ecological as well as economic point of view – than conventional greenhouse agriculture. The integration of insights can also yield new interventions such as new technologies. In this project, the students built an experimental fogponic system geared to their insights and research question, using plastic pipework. Nutrients were added to distilled water in the necessary quantities before entering the system. The water was subsequently led through the system to the dispenser. This dispenser used ultrasonic oscillations to turn the water into fog, which was led through the tubes in which the roots of the tomato plants were hanging so that they could absorb the nutrients. The remaining fog condensed and returned to the dispenser to ensure continuous circulation.

12 Interdisciplinary careers

In the previous chapters, we discussed what interdisciplinary research entails, when it can be applied, and why it should be applied. Throughout this book, we have seen that interdisciplinary research is by its very nature determined by the particular research questions and context of that research project – more so than in disciplinary research. This also affects the backgrounds, careers, and activities of interdisciplinarians themselves. Indeed, by looking at the biographies and statements of a few interdisciplinarians, you may recognize this context dependence as well as the pluralism in terms of scientific contents and methods that are challenging and inspiring them.

In this final chapter, we introduce four interdisciplinary researchers who will share their experiences with you and give tips on what has helped them in their academic careers. Below, they sketch their careers: how they became involved in interdisciplinary research, their current research endeavors, and their motivation and challenges in interdisciplinary research.

Prof. dr. Catherine Lyall
Professor of Science and Public Policy,
University of Edinburgh

Catherine Lyall has a bachelor's degree in chemistry and a master's degree in science and technology policy. She completed her PhD when she joined the University of Edinburgh as a part-time Research Development Officer, having worked previously in scientific publishing and for the UK civil service and a UK learned academy. Professor Lyall has worked on the EU-funded SHAPE-ID (Shaping interdisciplinary practices in Europe) and wrote a book titled *Being an Interdisciplinary Academic: How Institutions Shape University Careers* on the importance and misunderstandings of interdisciplinarity in academia.

Dr. Anna Beukenhorst
Strategic lead at Leyden Labs
and visiting scientist at the Harvard School of Public Health

Anna Beukenhorst specialized in data science methods for converting sensor data from smartphones and smartwatches into health insights. After a Bachelor of Science degree in Beta-Gamma Studies and a Master of Science in Medical Informatics, she completed a PhD in Medicine & Health Data Science and a Postdoc at the Harvard Department of Biostatistics. Currently, she is the strategic lead at the biotech startup Leyden Labs. She has won various prizes for essays on science technology and statistics.

Dr. Bianca Vienni Baptista
Senior researcher and lecturer of the Transdisciplinarity Lab, ETH Zürich

Bianca Vienni Baptista works in the field of Science, Technology and Society (STS) Studies, focusing in particular on the study of interdisciplinary and transdisciplinary knowledge production processes. As a result, she focuses her research on the specific conditions for interdisciplinary and transdisciplinary research and on the production and social use of knowledge in different countries, including the role of universities and other institutions. She has worked on the EU-funded SHAPE-ID (Shaping interdisciplinary practices in Europe) and is currently the group leader and principal investigator of the project *Investigating interdisciplinarity and transdisciplinarity: intersections of practices, culture(s) and policy in collaborative knowledge productions*. She is also part of an interdisciplinary academic community working in the field of cultural studies of science.

Dr. Jaco de Swart
Postdoctoral fellow at Science Technology and Society, Massachusetts Institute of Technology

Jaco de Swart, Helleman Postdoctoral Fellow at the Massachusetts Institute of Technology (MIT), finished his PhD in the history and philosophy of physics at University of Amsterdam (UvA) and has been a visiting researcher at the Institute for Advanced Study at Princeton University. He plays bass in a semi-professional rock band and owns three chickens.

How did you get involved in interdisciplinary research?

Catherine Lyall: 'I worked on a series of fixed-term, grant-funded, part-time research contracts in research centers linked to the School of Social and Political Science for 14 years before I was awarded my personal chair in 2013. I became interested in "interdisciplinarity" but did not reflect on the implications of that mode of research, especially on academic careers. I would credit Barry et al. (2008) and the Swiss td-net and their series of international conferences for showing me that this could become a career path.'

Anna Beukenhorst: 'From Dutch sign language to superpower China to statistics, I have always had a broad academic interest. My undergraduate degree in Liberal Arts and Sciences was like a candy shop with the possibility to follow all these courses and more (sustainability, the adolescent brain, big history, and a minor in programming). The interdisciplinary field of healthcare informatics and epidemiology turned out to make me tick: it combines programming, statistics, human behavior, medicine, and public health.'

Bianca Vienni Baptista: 'Since my PhD, I have built my career applying an anthropological perspective to connect Science and Technology Studies (STS) questions to interdisciplinary and transdisciplinary research (IDR/TDR) challenges. In my PhD, I applied an interdisciplinary approach to the socialization of knowledge in Uruguay. Then, I proposed an area of research entitled "Studies of Inter- and Transdisciplinarity" (SoIT) (Vienni-Baptista, 2016; Vienni-Baptista et al., 2020) within the field of STS to advance the construction of inter- and transdisciplinary institutional spaces in Latin America.'

Jaco de Swart: 'By trying to understand what is *behind* the topics and subjects that I studied. During my bachelor's, it appeared that I was not interested in just science, I also wanted to know why science is the way it is. How does science *itself* work? Why do we believe in things like black holes and microbes, and where do these ideas come from? These questions naturally tend to cross disciplinary borders.'

Can you tell us about your current job and how it is inter-/transdisciplinary?

Catherine Lyall: 'Like many academics, my job is a mix of research, teaching, and a lot of admin! Fostering interdisciplinarity has become a cornerstone of research policy in the UK and Europe, and my job also involves contributing to research policy through, for example, peer review and expert advice to research funders and other research groups. I have particularly enjoyed contributing to the EU SHAPE-ID project which has been advising the European Commission on how to better support inter- and transdisciplinary research and has produced a toolkit (see QR code) to help policymakers, researchers, and research partners make better decisions and promote change in these areas.'

Anna Beukenhorst: 'Smartphones are a game-changer for healthcare research. We carry them around wherever we go, and their sensors record our location, step count, phone use, the local weather, and air pollution. At Harvard, we developed an open-source smartphone app and analysis pipeline for medical research. We collaborate with hospitals. Smartphones enable us to test new medicines for neurodegenerative diseases without requiring wheelchair-based participants to come to the hospital weekly. It allows us to follow adolescents that are at high risk of self-harming or suicidal behavior. Or to alert someone that the recovery after their hernia operations is not going as planned: it turns out a smartphone can detect that long before the next doctor's visit. All these studies require knowledge of medicine (what do we need to measure?), on human behavior (how do we convince people to keep using our app?), on smartphones (how accurate is location data?), and on data science (is a step count algorithm valid on wheelchair users?).'

Bianca Vienni Baptista: 'Currently, I am finishing my work in a H2020 project *Shaping Interdisciplinary practices in Europe* (SHAPE-ID). The project examined best and poor practices in Arts, Humanities and Social Sciences integration in IDR/TDR in Europe. SHAPE-ID ultimately delivered a toolkit and recommendations (see QR code) to guide European policymakers, funders, universities, and researchers in achieving successful pathways to inter- and transdisciplinary integration.

As of January 2022, I am Group Leader and Principal Investigator of the project *Investigating interdisciplinarity and transdisciplinarity: intersections of practices, culture(s) and policy in collaborative knowledge productions*, funded by the Swiss National Science Foundation. The new research group is part of an interdisciplinary academic community working in cultural studies of science.

This project studies interdisciplinary and transdisciplinary research as a means of achieving transformative change for solving societal challenges. Many factors act as obstacles to high-impact collaborative research, resulting in gaps and disconnections in practices and policy. The main aim is to explore the intersections among practices (researchers), cultures (knowledge), and policy (institutions) in order to improve the responsiveness of interdisciplinary and transdisciplinary research to address scientific and societal challenges.'

Jaco de Swart: 'I combine ideas and methods from philosophy, history, and anthropology to study what scientists do when they study the universe – and how this has changed in the last 50 years. In that sense, I am a historian of the history of the universe.'

If you went from a disciplinary to an interdisciplinary career: what are the gains and what are the losses of going the interdisciplinary route in your research?

Catherine Lyall: "I enjoy the practical, "real world" aspects of work to influence policy change but the value of such work is not always recognized in academic institutions that attach greater importance to more theoretical contributions to knowledge. So, it has taken longer to become established in an academic career and I have no doubt that I could have gained a permanent university position sooner – and most probably had a stronger sense of "belonging" – if I had followed a more traditional route. These sentiments were echoed by many respondents with whom I conducted career history interviews for my recent book '*Being an Interdisciplinary Academic*'."

Bianca Vienni Baptista: 'Since my PhD, I have applied an interdisciplinary approach to my research, as I am interested in STS connections with IDR/TDR. Inter- and transdisciplinary research allows science to be more democratic and pluralistic. Although this is a gain, researchers embarking on inter-and transdisciplinary careers have a double identity. On the one hand, they (we) need to legitimize their work in their own field of knowledge, to have a "home" and to get funding. On the other hand, they (we) belong to a growing community of interdisciplinarians and transdisciplinarians and need to develop skills and expertise in collaborative research settings.'

What is your main motivation/inspiration for your current research interest?

Catherine Lyall: 'Working with Bianca Vienni Baptista (ETH Zürich) and Isabel Fletcher (University of Edinburgh), my current project is to publish an anthology of key readings in IDR/TDR. The goal of this edited collection is to provide new entrants with foundational knowledge in order to lower entry barriers, improve the conduct of IDR/TDR, and generally make it a bit less of a struggle so that every new researcher in this field doesn't have to "reinvent the wheel".'

Anna Beukenhorst: 'After my postdoc I wanted to go from observing the world through research into influencing the world through research. That's why I joined Leyden Labs, a start-up aiming at developing nasal sprays that provide immediate protection against pandemic viruses. Pandemic preparedness is very interdisciplinary: from tiny viral particles to the world's population dynamics, from policy to biotechnology. My work still revolves around research and interpreting data, and I'm enjoying the high-paced start-up environment.'

Bianca Vienni Baptista: 'I am contributing to improved policies for IDR/TDR and education that are grounded in aggregated evidence from different fields. These outcomes will enhance our understanding of rapidly changing conditions of academic work, putting IDR/TDR in the forefront of international discussions of science and technology.'

Jaco de Swart: 'To not take things for granted. I enjoy showing how things that *seem* necessary to us do not always have to be that way – whether it is a certain societal norm or a way of thinking about the origin of the cosmos.'

Do you have a tip for aspiring interdisciplinary and transdisciplinary researchers?

Catherine Lyall: 'My top tip would be to find allies and mentors. I would recommend the mentoring guide that we have produced as part of the SHAPE-ID toolkit. It has taken me longer to find my "invisible college" of peers than it would have done if I had pursued research within a single discipline. Friends and colleagues in networks such as td-net, i2S, AIS, and SHAPE-ID have helped to validate my work and give it international visibility, which is important when having a wider variety of research interests and projects inevitably leads to shifting networks of peers.'

Anna Beukenhorst: 'As an interdisciplinary student, I sometimes worried that my interests were too broad. From sustainable city gardens to doing business in China to clinical practice guidelines, I did projects on everything. In hindsight, those activities all helped shape my identity. And sometimes, even if I didn't end up in that field, they led to unexpected results: the sustainable city gardens project led to an amazing student side job, and an assignment on clinical practice guidelines eventually led to a PhD project and scholarship.'

Bianca Vienni Baptista: 'To persevere. The community is currently growing, and efforts to professionalize the field are emerging. Although a great deal of work is still necessary, networks and communities are being built and consolidated (the newly ITD alliance is a good example of this). More high-quality collaborative research is critically needed around the world, together with more in-depth knowledge on how and where to perform IDR/TDR and how to support them. For this, early career researchers with new perspectives and motivations are required.'

Jaco de Swart: 'Take your time to stop and think, look around, and be inspired. Work hard at the right moments, but do not work *too* hard! Overworking is dangerous and will also severely limit your passion and creativity.'

What do you see as the biggest challenge and/or promise in interdisciplinary research in the next ten years?

Catherine Lyall: 'From a UK perspective, Dutch universities in Amsterdam and Utrecht have been pioneers in curriculum-based interdisciplinary teaching but, in general, the institutionalization of interdisciplinary teaching needs to catch up with inter- and transdisciplinary research within most universities in Europe. Not least, this needs to happen so that we have an academic workforce better able to undertake IDR/TDR. Relatedly, I also see great advantages in promoting more cross-sector career moves so that universities can benefit from person-embodied, outside, "real world" knowledge and vice versa.'

Anna Beukenhorst: 'We need more interdisciplinary students to better shape our future with technology. "People worry that computers will get too smart and take over the world, but the real problem is that they're too stupid and they've already taken over the world." This quote from Pedro Domingos illustrates many of the most pressing problems in (bio)technology.

Interdisciplinary researchers are pivotal to solve these problems. What kind of relation do we want with technology? How can we preserve our privacy? How should we regulate black-box algorithms?'

Bianca Vienni Baptista: 'Even though policymakers and funding organizations are increasingly aiming at interdisciplinary and transdisciplinary collaboration, the corresponding methodologies and theories are not mainstream topics in either the academic or policy realms.

As we argued in a recent policy brief (Vienna Baptista et al., 2020): "Inter- and transdisciplinary careers are still seen as risky for researchers. Policymakers should support universities to build capacity in IDR and TDR by taking steps to de-risk inter- and transdisciplinary career paths and integrating IDR/TDR into education and training at an early stage."'

Jaco de Swart: 'The current norms and funding infrastructure in science still reward monodisciplinary thinking, unhealthy workloads, and absurd specialization. We need new creative ways to think differently about careers in science! And the biggest promise? You!'

Further reading

Now that you have experienced the interdisciplinary research process, you are probably eager to find out more about interdisciplinary research in the academic world. Interdisciplinary research can be found in many different forms and places; the following books and organizations are good places to start.

I Books on interdisciplinary research

Interdisciplinary Research: Process and Theory
Allen Repko, Rick Szostak
Sage Publications, 2017 (3rd edition)
This book is written for advanced undergraduate and graduate students and covers interdisciplinary research methods. It neatly describes how to achieve, produce, and express integration. This book can be very useful when writing an interdisciplinary research paper.

Introduction to Interdisciplinary Studies
Allen Repko, Rick Szostak, Michelle Phillips Buchberger
Sage Publications, 2019 (3rd edition)
This book is an introduction to the principles of interdisciplinary studies and is a helpful guide for those working with topics, problems, or themes that span multiple disciplines.

Case Studies in Interdisciplinary Research
Allen Repko, William Newell, Rick Szostak (eds.)
Sage Publications, 2011
Written by leading researchers in interdisciplinary research, these case studies show how to apply the interdisciplinary research process to a variety of problems.

Interdisciplinary Research Journeys
Catherine Lyall, Ann Bruce, Joyce Tait, Laura Meagher
Financial Times Press, 2011
This book provides a practical guide for researchers and research managers who are seeking to develop interdisciplinary research strategies at a personal, institutional, and multi-institutional level.

Being an Interdisciplinary Academic: How Institutions Shape University Careers
Catherine Lyall
Palgrave, 2019
This book highlights the importance of interdisciplinarity in the academic landscape and examines how it is understood in the context of the modern university. It addresses these issues and interdisciplinary academic careers on both a personal and systemic level, identifying how a resilient researcher can craft their own research trajectory to view interdisciplinarity as a truly embedded approach.

Interdisciplinary and Transdisciplinary Failures: Lessons Learned from Cautionary Tales
Dena Fam, Michael O'Rourke (eds.)
Routledge, 2021
This book provides insight for interdisciplinary and transdisciplinary researchers and practitioners on what *doesn't* work. Documenting detailed case studies of project failure if important, not only as an illustration of experienced challenges but also because projects do not always follow step-by-step protocols of preconceived and theorized processes.

Interdisciplinary Research: Case Studies from Health and Social Science
Frank Kessel, Patricia Rosenfield, Norman Anderson (eds.)
Oxford University Press, 2008
Although the case studies in this volume all stem from health and social sciences, their extensive analysis is illuminating for students from other fields as well.

The Toolbox Dialogue Initiative: The Power of Cross-Disciplinary Practice
Graham Hubbs, Michael O'Rourke, Steven Hecht Orzack (eds.)
CRC Press, 2020
Inspired by Toolbox Dialogue Initiative, this book presents dialogue-based methods designed to increase mutual understanding among collaborators in order to enhance the quality and productivity of cross-disciplinary collaboration. It also provides readers with a theoretical context, principal activities, and evidence for effectiveness.

Issues in Interdisciplinary Studies (journal)
Issues in Interdisciplinary Studies, founded in 1982, is an international, peer-reviewed publication of the Association for Interdisciplinary Studies dedicated to advancing the theory and practice of the many varieties of interdisciplinarity in academics and in the society at large. Published by Texas Tech University Press, its back issues are available at https://interdisciplinarystudies.org/issues/

Beyond Interdisciplinarity. Boundary Work, Communication, and Collaboration
Julie Thompson Klein
Oxford University Press, 2021
After decades of contributing to and even shaping the field, Julie Thompson Klein depicts in this book the heterogeneity and boundary work of interdisciplinarity and transdisciplinarity in a conceptual framework based on an ecology of spatializing practices in transaction spaces, including trading zones and communities of practice. The book includes both 'cross-disciplinary' work (encompassing multidisciplinary, interdisciplinary, and transdisciplinary forms) as well as 'cross-sector' work (spanning disciplines, fields, professions, government and industry, and communities).

Strategies for Team Science Success: Handbook of Evidence-Based Principles for Cross-Disciplinary Science and Practical Lessons Learned from Health Researchers
Kara L. Hall, Amanda L. Vogel, Robert T. Croyle
Springer, 2019
With contributions from more than 100 experts from a wide range of disciplines and professions, this volume offers readers a comprehensive set of actionable strategies for reducing barriers and overcoming challenges and includes practical guidance on how to implement effective team science practices.

Methods for Transdisciplinary Research
Matthias Bergmann, Thomas Jahn, Tobias Knobloch, Wolfgang Krohn, Christian Pohl, Engelbert Schramm (eds.)
Campus Verlag GmbH, 2013
This book provides scholars with a model for conceptualizing and executing transdisciplinary research, while offering a systematic description of methods for knowledge integration that can be applied to any field of research.

Enhancing Communication & Collaboration in Interdisciplinary Research
Michael O'Rourke, Stephen Crowley, Sandford Eigenbrode, Jeffrey Wulfhorst (eds.)
Sage Publication, 2013
The book contains theoretical perspectives, case studies, communication tools, and institutional perspectives of interdisciplinary research.

Seeing the City. Interdisciplinary Perspectives on the Study of the Urban
Nanke Verloo, Luca Bertolini
Amsterdam University Press, 2020
A rich volume that embraces a variety of perspectives and provides an essential collection of methodologies for studying the city from multiple, interdisciplinary, and transdisciplinary perspectives.

The Oxford Handbook of Interdisciplinarity
Robert Frodeman, Julie Thompson Klein, Roberto Carlos Dos Santos Pacheco (eds.)
Oxford University Press, 2017 (2nd edition)
This handbook provides a synoptic overview of the current state of interdisciplinary research, education, administration, management, and problem-solving knowledge that spans the disciplines and interdisciplinary fields and crosses the space between the academic community and the society at large.

Facilitating Interdisciplinary Research
National Academy of Sciences, 2004
The report identifies steps that researchers, teachers, students, institutions, funding organizations, and disciplinary societies can take to conduct, facilitate, and evaluate interdisciplinary research programs and projects more effectively. Throughout the report, key concepts are illustrated with case studies and results of the committee's surveys of individual researchers and universities provosts.

Interdisciplinarity: Essays from Literature
William Newell (ed.)
College Entrance Examination Board, 1998
This is an interesting collection of essays on various aspects of interdisciplinarity, ranging from administration to interdisciplinarity in different scientific domains.

Action Research Plus Foundation
The foundation's aim is to make global knowledge democracy more available by supporting inter/transdisciplinary dialogue and connectivity for those practicing at the developmental edge of action research worldwide.
https://actionresearchplus.com

The Association of Interdisciplinary Studies (AIS)
The AIS is the main interdisciplinary professional organization founded in 1979 to promote the interchange of ideas among scholars and administrators in all of the arts and sciences regarding interdisciplinarity and integration. The AIS journal Issues in Interdisciplinary Studies (formerly, Issues in Integrative Studies) publishes peer reviewed articles on a wide range of interdisciplinary topics, research, education and policy. (Editor and co-author Keestra is currently President of AIS.)
https://interdisciplinarystudies.org/

Center for Interdisciplinary Research (ZiF)
ZiF was established in 1968 at the new Bielefeld University. It houses and funds interdisciplinary research projects in the natural sciences, humanities, and social sciences.
http://www.uni-bielefeld.de/ZIF/

Global Inter- and Transdisciplinary Alliance
Bringing together various associations and networks, the purpose of the Global Alliance for Inter- and Transdisciplinarity (ITD Alliance) is to strengthen and to promote the global capacity and the calibre of collaborative modes of boundary-crossing research and practive.
https://itd-alliance.org

International Society of Political Psychology (ISPP)
The ISPP is an interdisciplinary organization representing all fields of inquiry concerned with exploring the relationships between political and psychological processes. Members include psychologists, political scientists, psychiatrists, historians, sociologists, economists, and anthropologists, as well as journalists, government officials, and others.
http://www.ispp.org/

The Resilience Alliance (RA)
The RA is a research organization of scientists and practitioners from many disciplines who collaborate to explore the dynamics of social–ecological systems. The body of knowledge developed by the RA encompasses key concepts of resilience, adaptability, and transformability, and provides a foundation for sustainable development policy and practice.
http://www.resalliance.org/

The Santa Fe Institute
The Santa Fe Institute is a private, not-for-profit, independent research and education center, founded in 1984, where leading scientists grapple with some of the most compelling and complex problems of our time. Researchers come to the Santa Fe Institute from universities, government agencies, research institutes, and private industry to collaborate across disciplines, merging ideas and principles of many fields – from physics, mathematics, and biology, to the social sciences and the humanities.
http://www.santafe.edu/

Society for the Advancement of Socio-Economics (SASE)
The SASE is an international, interdisciplinary academic organization. The academic disciplines represented in the SASE include economics, sociology, political science, organization studies, management, psychology, law, and history.
https://sase.org/

Td-net
Td-net is a network initiated by the Swiss academy of sciences and advances the mutual learning between interdisciplinary and transdisciplinary researchers and lecturers across thematic fields, languages, and countries, thereby supporting community building.
http://www.transdisciplinarity.ch

Index

Action research 69-70, 86
Actionability 70, 86
AIS 147, 154
Assumptions 21, 35, 106-109, 120
Axiomatics 44

Boundaries 43-46, 124
Boundary object 86

Careers 142-149
Case study 39, 57, 93
Collaboration 50-51, 95-98
Common ground 97, 113-118
Community of scientists 46
Complex Adaptive System 62
Complexity, complex problem
 39, 59, 62
Concepts 71, 107-109, 122, 128
Conclusion 132
Consciousness 26, 36, 63, 76

Data
 Analysis 127-128
 Collection 61, 126-127
 Management table 108, 124
 Qualitative 36, 95, 123-127
 Quantitative 78, 95, 123-127
Deduction 28
Dialogue methods 88
Dimensions 55, 122
Discipline 42, 107
 Disciplinary perspectives
  18, 31, 53, 113, 131
Discussion 129

Emergent properties 51, 59
Empirical cycle 23
Ethics 44, 61
Evaluation 25, 32
Experimentation 45
Explanation 54
Extra-academic stakeholders 53, 66

Facts 24
Framework 26, 76

Gap of knowledge 23, 131
Gap of reasoning 131

Hermeneutic method 33
Heuristics 76
Hypothesis 32, 108
i2S 147
Implementation 86
Indicators 122, 128
Induction 25, 45
Insight into the problem 109
Instruments 32, 45, 78
Integration
 19, 43, 53, 71-89, 116, 125, 130
Interdisciplinarity 42, 50, 52, 71, 75
 Interdisciplinary research
  model 91
Interpret results 130, 141
Intervention 17, 53, 64
Isolation 48
Iterative, recursive 75, 92, 119

Knowledge 15, 27, 38, 41, 42-46

Laws 24-28

Mechanistic 85, 100, 101
Methods 41, 78-80, 94
 Methodological framework
  121-127
Mind map 100

Model			Science	
Animal	80		Cycle	24-42, 71-72
Computational	81		Of Team Science	75, 87-88
Explanatory	83		SHAPE-ID	145-147
Simulation	80		Specialization	48-50
Climate change	84		Sub-question	119
Multi-				
Causal	26		Td-net	75, 144, 155
Disciplinarity	52-54		Teamwork: *See* Collaboration	
Interpretability	65		Techniques, integration	
Level network	63, 64		techniques	71-89
			Theoretical	
Observations	32-34		Analysis	71-72, 93, 105-120
Operationalization	31-33		Framework	105-120
Optimization function	81		Theory-ladenness	32
Orientation	93, 99		Toolbox dialogue method	88
Overlap	103, 105, 112, 137		Transdisciplinarity	52-54, 66-69
			Triangulation	31-36, 99, 103
Paradigm	46, 47, 115			
Perception	25, 39, 45		Unification of science	50-52
Perspectives	18-20, 31, 36, 43, 73, 100, 107, 112		Universities	46, 54
Philosophy of science	21-42		Variables	80-81, 122, 128
Pluralism			Visualization	130
Integrative	74			
Methodological	35-41, 73-78, 113-115		Wicked problem	65-69
Theoretical	35-41, 76-77			
Pre-disciplinary	43-45			
Predictions	28-31			
Reasoning	22-30, 34-41			
Reductionism	50-51			
Reflection	23-24, 87-88, 95, 133			
Relevance	56-57, 67-68			
Research				
Approach	117, 123-124			
Design	122-126			
Phases	92-95			
Question	101-103, 119			
Results	119-128			
Revolution	45, 61			
Robustness	34-36, 64, 67			

References

Apostel, L., Berger, G., Briggs, A., & Machaud, G. (1972). *Interdisciplinarity: Problems of teaching and research at universities*. Paris: Organization for Economic Cooperation and Development.

Bammer, G. (2013). *Disciplining interdisciplinarity: Integration and implementation sciences for researching complex real-world problems*. Canberra: ANU Press.

——, D. McDonald, and P. Deane (2009). *Research integration using dialogue methods*. Canberra: ANU Press.

——, M. O'Rourke, D. O'Connell, L. Neuhauser, G. Midgley, J.T. Klein, G.P. Richardson, et al. (2020). 'Expertise in research integration and implementation for tackling complex problems: When is it needed, where can it be found and how can it be strengthened?'. *Palgrave Communications, 6*(5).

Barry, A., G. Born, and G. Weszkalnys (2008). 'Logics of interdisciplinarity'. *Economy and Society, 37*(1) 20-49.

Bechtel, W., and R.C. Richardson (2010). *Discovering complexity: Decomposition and localization as strategies in scientific research*. Cambridge, MA: MIT Press.

Benjamin, D.J., J.O. Berger, M. Johannesson, B.A. Nosek, E.J. Wagenmakers, R. Berk, V.E. Johnson et al. (2018). 'Redefine statistical significance'. *Nature Human Behaviour, 2*(1) 6-10.

Bergmann, M., T. Jahn, T. Knobloch, W. Krohn, C. Pohl, and R.C. Faust (2012). *Methods for transdisciplinary research: A primer for practice*. Frankfurt: Campus Verlag.

Berkes, F. (2003). 'Alternatives to conventional management: Lessons from small-scale fisheries'. *Environments, 31*(1) 5-19.

Bocci, C., and P. Freguglia (2006). 'The geometric side for an axiomatic theory of evolution'. *Rivista Di Biologia - Biology Forum, 99*(2). 307-325.

Borda, O.F. (1979). 'Investigating reality in order to transform it: The Colombian experience'. *Dialectical Anthropology, 4*(1) 33-55.

Boyer, P., and P. Liénard (2006). 'Why ritualized behavior? Precaution systems and action parsing in developmental, pathological and cultural rituals'. *Behavioral and Brain Sciences, 29*(6) 595-613.

Burkert, W. (1985). *Greek religion*. Cambridge, MA: Harvard University Press.

Caetano, R., J. Schafer, and C.B. Cunradi (2001). 'Alcohol-related intimate partner violence among White, Black, and Hispanic couples in the United States'. *Alcohol Research and Health, 25*(1) 58-65.

Calhoun, C., and D. Rhoten (2010). 'Integrating the social sciences: Theoretical knowledge, methodological tools, and practical applications'. In R. Frodeman, J. Klein, C. Mitcham, and J. Holbrook (eds.), *The Oxford Handbook of Interdisciplinarity* (pp. 103-118). Oxford: Oxford University Press.

Cao, J., T.J. Cogdell, D.F. Coker, H. Duan, J. Hauer, U. Kleinekathöfer, D. Zigmantas, et al. (2020). 'Quantum biology revisited'. *Science Advances,* (6).

Charles, A.T. (1994). 'Towards sustainability: The fishery experience'. *Ecological Economics*, 11(3) 201-211.

Chris, L., S. Donovan, M. O'Rourke, S. Crowley, S.D. Eigenbrode, L. Rotschy, J.D. Wulfhorst, et al. (2016). 'Seeing through the eyes of collaborators: Using toolbox workshops to enhance cross-disciplinary communication'. In M. O'Rourke, S. Crowley, S.D. Eigenbrode, J.D. Wulfhorst (eds.) *Enhancing communication & collaboration in interdisciplinary research*. Thousand Oaks: SAGE Publications, Inc.

Churchland, P.S. (2005). 'A neurophilosophical slant on consciousness research'. *Progress in Brain Research, 149*, 285-293.

Cohen, H.F. (1994). *The scientific revolution: A historiographical inquiry*. Chicago: University of Chicago Press.

Craver, C.F. (2007). *Explaining the brain mechanisms and the mosaic unity of neuroscience*. Oxford: Oxford University Press.

Darden, L., and N. Maull (1977). 'Interfield theories'. *Philosophy of Science, 44*(1), 43-64.

Dautenhahn, K., C.L. Nehaniv, M.L. Walters, B. Robins, H. Kose-Bagci, N.A. Mirza, and M. Blow (2009). 'KASPAR – a minimally expressive humanoid robot for human-robot interaction research'. *Applied Bionics and Biomechanics, 6*(3-4), 369-397.

De Soto, H. (2003). *The mystery of capital: Why capitalism triumphs in the West and fails everywhere else*. New York: Basic Books.

Diener, E., R.A. Emmons, R.J. Larsem, and S. Griffin (1985). 'The satisfaction with life scale'. *Journal of Personality Assessment, 49*(1), 71-75.

European Commission. (2004). 'The role of the universities in the Europe of knowledge'. *European Education, 36*(2), 5-34.

Fish, E.W., S. Faccidomo, and K.A. Miczek (1999). 'Aggression heightened by alcohol or social instigation in mice: Reduction by the 5-HT(1B) receptor agonist CP-94,253'. *Psychopharmacology, 146*(4), 391-399.

Flyvbjerg, B. (2006). 'Five misunderstandings about case-study research'. *Qualitative Inquiry, 12*(2) 219-245.

Frank, R. (1988). ''Interdisciplinary': The first half century'. *Issues in Integrative Studies, 6*, 91-101.

Freire, P. (2005). *Pedagogy of the oppressed: 30th anniversary edition*. Trans. M. Bergman Ramos. New York: Continuum.

Frodeman, R. (2013). *Sustainable knowledge: A theory of interdisciplinarity*. London: Palgrave Pivot.

Fuller, S. (2017). 'The military-industrial route to interdisciplinarity'. In Frodeman, R., *The Oxford Handbook of Interdisciplinarity*. Oxford: Oxford University Press.

Garcia, S.M., and K.L. Cochrane (2005). 'Ecosystem approach to fisheries: A review of implementation guidelines'. *ICES Journal of Marine Science, 62*(3), 311-318.

Hedström, P., and R. Swedberg (1998). *Social mechanisms: An analytical approach to social theory*. Cambridge: Cambridge University Press.

Heffernan, O. (2010). 'Earth science: The climate machine'. *Nature, 463*, 1014-1016.

Heinz, A.J., A. Beck, A. Meyer-Lindenberg, P. Sterzer, and A. Heinz (2011). 'Cognitive and neurobiological mechanisms of alcohol-related aggression'. *Nature Reviews, Neurosciences, 12*(7), 400-413.

Henrich, J. (2020). *The WEIRDest people in the world*. New York: Farrar, Straus and Giroux.

Hindenlang, K.E., J. Heeb, and M. Roux (2008). 'Sustainable coexistence of ungulates and trees: A stakeholder platform for resource use negotiations'. In G. Hirsch Hadorn, H. Hoffman-Riem, S. Biber-Klemm, W. Grossenbacher-Mansuy, D. Joye, C. Pohl, U. Wiesmann, and E. Zemp (eds.), *Handbook of transdisciplinary research*. New York: Springer.

Hirsch Hadorn, G., C. Pohl, H. Hoffmann-Riem, S. Biber-Klemm, W. Grossenbacher-Mansuy, D. Joye, C. Pohl, U. Wiesmann, and E. Zemp (eds.) (2008). *Handbook of transdisciplinary research*. New York: Springer.

Holland, J.H. (2014). *Complexity: A very short introduction*. Oxford: Oxford University Press.

Jakeman, A.J., R.A. Letcher, and J.P. Norton (2006). 'Ten iterative steps in development and evaluation of environmental models'. *Environmental Modelling and Software, 21*(5), 1-13.

Keestra, M. (2000). 'Aristoteles'. In M. Keestra (ed.), *Tien westerse filosofen*. Amsterdam: Nieuwezijds.

Keestra, M. (2012). 'Understanding human action: integrating meanings, mechanisms, causes, and contexts'. In A. Repko, W. Newell and R. Szostak (eds.), *Case Studies in Interdisciplinary Research*. Thousand Oaks, CA: Sage Publications.

——. (2017). 'Metacognition and reflection by interdisciplinary experts: Insights from cognitive science and philosophy'. *Issues in Interdisciplinary Studies, 35*, 121-169.

——. (2019). Imagination and actionability: Reflections on the future of interdisciplinarity, inspired by Julie Thompson Klein. *Issues in Interdisciplinary Studies, 37*(2), 110-129.

——, and S.J. Cowley (2009). 'Foundationalism and neuroscience; silence and language'. *Language Sciences, 31*(4), 531-552.

Klein, J.T. (1990). *Interdisciplinarity: History, theory and practice*. Detroit, MI: Wayne State University Press.

———, and W.H. Newell (1997). 'Advancing interdisciplinary studies.' In J. Gaff and J. Ratcliff (eds.), *Handbook of the undergraduate curriculum: Comprehensive guide to purposes, structures, practices, and change* (1st ed.) San Francisco, CA: Jossey-Bass.

——— (2004). 'Prospects for transdisciplinarity'. *Futures, 36*(4), 515-526.

——— (2017). 'Typologies of interdisciplinarity: The boundary work of definition'. In R. Frodeman, *The Oxford Handbook of Interdisciplinarity*. Oxford: Oxford University Press.

Krohn, W. (2017). 'Interdisciplinary cases and disciplinary knowledge: Epistemic challenges of interdisciplinary research'. In R. Frodeman, *The Oxford Handbook of Interdisciplinarity*. Oxford: Oxford University Press.

Kuhn, T.S. (1970). *The structure of scientific revolutions* (2nd ed.). Chicago: University of Chicago Press.

Lafferty, P., and J. Rowe (1997). *Dictionary of science*. London: Brockhampton Press.

Laudan, L. (1986). *Science and values – the aims of science and their role in scientific debate*. Oakland: University of California Press.

Le Treut, H., R. Somerville, U. Cubasch, Y. Ding, C. Mauritzen, A. Mokssit, M. Prather, et al. (2007). 'Chapter 1: Historical overview of climate change science'. *Climate Change 2007: The Physical Science Basis. Contribution of Working Group I to the Fourth Assessment Report of the Intergovernmental Panel on Climate Change*.

LeDoux, J.E. (2014). 'Coming to terms with fear'. *Proceedings of the National Academy of Sciences, 111*(8), 2871-2878.

Lélé, S., and R.B. Norgaard (2005). 'Practicing interdisciplinarity'. *Bioscience, 55*(11), 967-975.

Long, P.O., and H.F. Cohen (1996). *The scientific revolution: A historiographical inquiry*. Chicago: Chicago University Press.

Looney, C., D. Shannon, M. O'Rourke, S. Crowley, S.D. Eigenbrode, L. Rotschy, J.D. Wulfhorst, et al. (2016). 'Seeing through the eyes of collaborators: Using toolbox workshops to enhance cross-disciplinary communication'. In S.D. Eigenbrode and J.D. Wulfhorst, (eds.) *Enhancing communication & collaboration in interdisciplinary research*. Thousand Oaks: SAGE Publications, Inc.

Lyall, C. (2008). *Designing interdisciplinary research for policy and practice*. Edinburgh: University of Edinburgh.

National Academy of Sciences, Engineering and Medicine. (2005). *Facilitating interdisciplinary research*. Washington D.C.

Newell, W.H. (2007). 'Decision making in interdisciplinary studies'. In G. Morçöl (ed.), *Handbook of decision making in interdisciplinary studies* (pp. 245-264). New York: CRC Press/Taylor & Francis Group.

—— (2013) The State of the Field: Interdisciplinary Theory. *Issues in Interdisciplinary Studies 31*, 22-43

Norder, S.J., and K.F. Rijsdijk (2016). 'Interdisciplinary island studies: Connecting the social sciences, natural sciences and humanities'. *Island Studies Journal, 11*(2), 673-686.

O'Rourke, M., S. Crowley, B. Laursen, B. Robinson, and S.E. Vasko (2019). 'Disciplinary diversity in teams: Integrative approaches from unidisciplinarity to transdisciplinarity'. In K.L. Hall, A.L. Vogel, and R.T. Croyle, *Strategies for team science success*. New York: Springer.

Oppenheim, P., and H. Putnam (1958). *Unity of science as a working hypothesis*. Minneapolis: University of Minnesota Press.

Page, S.E. (2010). *Diversity and complexity*. Princeton: Princeton University Press.

Paul, R., L. Elder, and Foundation for Critical Thinking (2012). 'The miniature guide to critical thinking: Concepts and tools'. *27th International Conference on Critical Thinking, Berkeley*.

Pohl, C., and G. Hirsch Hadorn (2007). *Principles for designing transdisciplinary research*. München: Oekom Verlag.

——, P. Krütli, and M. Stauffacher (2017). 'Ten reflective steps for rendering research societally relevant'. *Gaia, 26*(1), 43-51.

Popper, K.R. (2002). *Conjectures and refutations: The growth of scientific knowledge*. London: Routledge.

Reason, P., and H. Bradbury (2008). 'Introduction'. In P. Reason and H. Bradbury (eds.), *The SAGE handbook of action research* (pp. 1-10). London: SAGE Publications.

Repko, A.F., and R. Szostak (2017). *Interdisciplinary research: Process and theory*. Thousand Oaks: SAGE Publications.

—— (2008). Defining interdisciplinary studies. In A. Repko, *Interdisciplinary Research: Process and Theory*. Thousand Oaks: SAGE Publications.

Resnik, D.B. (1998). *The ethics of science: An introduction*. London: Routledge.

Rittel, H.W.J., and M.M. Webber (1973). 'Dilemmas in a general theory of planning'. *Policy Sciences, 4*(2) 155-169.

Ruben, D.H. (1992). *Explaining explanation*. New York: Routledge.

Shannon, C.E. (1948). 'A mathematical theory of communication'. *Bell System Technical Journal, 27*(3) 379-423.

Sousanis, N. (2015). *Unflattening*. Cambridge, MA: Harvard University Press.

Stichweh, R. (2001). Scientific disciplines, history of. In J.S. Neil and B.B. Paul (eds.), *International encyclopedia of the social & behavioral sciences* (pp. 13727-13731). Oxford: Pergamon.

Stramba-Badiale, M., K.M. Fox, S.G. Priori, P. Collins, C. Daly, I. Graham, M. Tendera, et al. (2006). 'Cardiovascular diseases in women: A statement from the policy conference of the European society of cardiology'. *European Heart Journal, 27*(8), 994-1005.

Szostak, R. (2004). *Classifying science: Phenomena, data, theory, method, practice*. Dordrecht: Springer.

Talisayon, V.M. (2010). *Development of scientific skills and values in physics education*.

Tse, M., H. Yu, N. Kijbunchoo, A. Fernandez-Galiana, P. Dupej, L. Barsotti, J.. Zweizig, et al. (2019). 'Quantum-enhanced advanced LIGO detectors in the era of gravitational-wave astronomy'. *Physical Review Letters, 123*(23), 231107-(1-8).

Vander Valk, F. (2013). *Essays on neuroscience and political theory: Thinking the body politic*. London: Routledge.

Van der Wal, J. M., C.D. Van Borkulo, M.K. Deserno, J.J.F. Breedvelt, M. Lees, J.C. Lokman, R.W. Wiers, et al. (2021). 'Advancing urban mental health research: From complexity science to actionable targets for intervention'. *The Lancet Psychiatry, 8*(11), 991-1000.

Vassli, L.T., and B.A. Farshchian (2018). 'Acceptance of health-related ICT among elderly people living in the community: A systematic review of qualitative evidence'. *International Journal of Human-Computer Interaction, 34*(2), 99-116.

Vienni Baptista, B. (2016). 'Among institutions, spaces and networks: Interdisciplinary and transdisciplinary realms in the American continent'. *INTERdisciplina, 4*(10), 22-34.

——, and S. Rojas-Castro (2020). 'Transdisciplinary institutionalization in higher education: A two-level analysis'. *Studies in Higher Education, 45*(6), 1075-1092.

Weingart, P. (2010). 'A short history of knowledge formations'. In R. Frodeman, J.T. Klein, and C. Mitcham (eds.), *The Oxford Handbook of Interdisciplinarity* (pp. 3-14). Oxford: Oxford University Press.

Wimsatt, W.C. (2007). *Re-engineering philosophy for limited beings*. Cambridge, MA: Harvard University Press.

References
Research projects undertaken by IIS students

Bakker, T., I. van der Linden, M. Steenbrink, M. Stuur, and S. Veldhuyzen van Zanten (2014). *Fogponics: Richting een duurzaam alternatief voor landbouw*, bachelor's thesis.

Bekius, F., and L. Elsenburg (2010). *De invloed van omega-3 vetzuren en meetmethode op HRV bij mannen en vrouwen*, bachelor's thesis.

Van Dierendonck, R.C.H., M.A.N.E. Van Egmond, D.L. Ten Hagen, and J. Kreuning (2013). *De dodo op de weegschaal. een verscherping van de regressiemethode ter bepaling van het gewicht van Raphus cucullatus*, bachelor's thesis.

Gellauf, M.N.A., W.L. Gravemaker, A.L. Isarin, and A.C. Waajen (2015). *Natriumsulfaatmist: Effect op lichtintensiteit, bodemsamenstelling en de groei van gewassen*, bachelor's thesis.

Schram, R. (2012). *Vrijwilligerstoerisme, een vloek of een zegen?*, bachelor's thesis.

Colophon

University of Amsterdam

The University of Amsterdam (UvA), with its 30,000 students, 5,000 staff, and a budget of more than 600 million euros, is one of the largest comprehensive universities in Europe. Teaching and research at the UvA are conducted at seven faculties: the Humanities, Social and Behavioral Sciences, Economics and Business, Law, Science, and Medicine and Dentistry, with programs offered in almost every field.

The Institute for Interdisciplinary Studies

The Institute for Interdisciplinary Studies (IIS) is a knowledge center for interdisciplinary learning and teaching. Each year, the institute provides a diversity of interdisciplinary education to some 3,300 students enrolled in bachelor's or master's programs or open courses. In recent years, dozens of lectures in different disciplines from within and outside the UvA have contributed to education or other activities at the institute. The IIS is a 'laboratory' for interdisciplinary experiments and projects that might lead to new interdisciplinary courses, teaching methods, or programs. The IIS conducts assignments and projects for clients both within and outside the UvA. It also advises on interdisciplinary education.

Authors

Dr. Machiel Keestra: Assistant professor of philosophy at the Institute for Interdisciplinary Studies, Faculty of Science, and Central Diversity Officer, University of Amsterdam. Past president of the international Association for Interdisciplinary Studies, founding board member of the global Inter- and Transdisciplinary Alliance, editorial board member of *Issues in Interdisciplinary Studies*.

Anne Uilhoorn, MSc.: Lecturer in Future Planet Studies at the Institute for Interdisciplinary Studies, Faculty of Science, University of Amsterdam.

Dr. ing. Jelle Zandveld: Former lecturer in Beta-Gamma Studies at the Institute for Interdisciplinary Studies, Faculty of Science, University of Amsterdam. Currently assistant professor at the Education Institute Biology at Utrecht University.

Contact
Institute for Interdisciplinary Studies
Science Park 904
1098 XH Amsterdam
Tel. +31 20 525 51 90
www.iis.uva.nl
Onderwijslab-iis@uva.nl